Level 1

CARPENTRY & JOINERY

NVQ/SVQ and CAA Diploma

carillion

www.pearsonschoolsandfecolleges.co.uk

✓ Free online support
✓ Useful weblinks
✓ 24 hour online ordering

0845 630 44 44

Heinemann

Part of Pearson

Heinemann is an imprint of Pearson Education Limited, a company incorporated in England and Wales, having its registered office at Edinburgh Gate, Harlow, Essex, CM20 2JE. Registered company number: 872828

www.pearsonschoolsandfecolleges.co.uk

Heinemann is a registered trademark of Pearson Education Limited

Text © Carillion Construction Ltd 2010

First published 2010

14 13 12 11 10
10 9 8 7 6 5 4 3 2 1

British Library Cataloguing in Publication Data is available from the British Library on request.

ISBN 978 0 435027 02 5

Typeset by Tek-Art, Crawley Down, West Sussex
Original illustrations © Pearson Education Ltd
Illustrated by Oxford Designers and Illustrators
Cover design by Wooden Ark
Printed in the UK by Scotprint

Carillion would like to thank the following people for their contribution to this book: Kevin Jarvis and John Harvie McLaughlin.

Pearson Education Limited would like to thank the following people for providing technical feedback: Graeme Coventry from Basingstoke College of Technology; Rob Harrison from Stoke College; and Jim Neil and Gerry McGowan from North Glasgow College.

Websites

The websites used in this book were correct and up-to-date at the time of publication. It is essential for tutors to preview each website before using it in class so as to ensure that the URL is still accurate, relevant and appropriate. We suggest that tutors bookmark useful websites and consider enabling students to access them through the school/college intranet.

The information and activities in this book have been prepared according to the standards reasonably to be expected of a competent trainer in the relevant subject matter. However, you should be aware that errors and omissions can be made and that different employers may adopt different standards and practices over time. Before doing any practical activity, you should always carry out your own Risk Assessment and make your own enquires and investigations into appropriate standards and practices to be observed.

Acknowledgements

Photo acknowledgements
The publisher would like to thank the following for their kind permission to reproduce their photographs:

(Key: b-bottom; c-centre; l-left; r-right; t-top)

Alamy Images: Charles Stirling 187/5.37, D Hurst 182/5.22, David J Green 62, 74, 191/5.52, 217/6.33, David Lawrence 131/3.39, Geoff du Feu 84, Image Source Pink 23, Justin Kase 29/1.19, 131/3.42, Nic Hamilton 3, Oleksiy Maksymenko 146/4.33, Peter Jordan 140/4.12, Robert Hunt 186/5.34, The Photolibrary Wales 25, Webstream 220/6.47; **Construction Photography**: Adrian Sherratt 129/3.32, 134, Buildpix 15, 131/3.38, Chris Henderson 7, 121, David Potter 53/1.40, 129/3.30, David Townsend 180/5.16, Ken Price 51/1.34, Sally-Ann Norman 51/1.35, Xavier de Canto 12; **Corbis**: 53/1.38; **CSCS**: 14; **John Gavin Photography**: 191/5.90; **Getty Images**: PhotoDisc 26, 52, 113, Photonica 175, Stone 201; **iStockphoto**: Ales Veluscek 165c, Bill Noll 165b, 166t, 166c, 167t, 167c/2, 167b, 168t, 168b, 174b, Dave White 164t, 174t, Dmitriy Pochitalin 166b, Ivan Vasilev 164c, Leigh Schindler 67/1.60, Nancy Nehring 165t, Tibor Nagy 167c/1; **Krys Bailey Marmotta PhotoArt** : 180/5.12; **Martyn F. Chillmaid**: 219/6.42; **PAT Training Services Limited**: 205/6.7; **Pearson Education Ltd**: Chris Honeywell 65/1.52, Clark Wiseman, Studio 8 221, David Sanderson 67/1.58, 67/1.59, 69/1.63, 69/1.64, 69/1.65, 70, Jules Selmes 19, Trevor Clifford 141/4.16; **Quayside Graphics**: 160; **Robert Clare**: 184/5.28; **Science Photo Library Ltd**: Gerry Watson 30l, Scott Camazine 24; **Shutterstock**: 7505811966 220/6.46, ABImages 117, Alex Kosev 65/1.51, Anatoly43 182/5.19, Andrey Bayda 30c, Dainis Derics 219/6.38, Dutsy Cline 220/6.45, Edd Westmacott 29cl, 66/1.56, Fekete Tibor 131/3.41, Frances A. Miller 51/1.33, Gilmanshin 75, Glue Stock 206, Guy Erwood 53/1.39, IOFoto 66/1.54, Jiri Hera 218/6.36, Joachim Wendler 139/4.9, Josef Bosak 217/6.29, Kevin Britland 130, 131/3.40, Martina Orlich 164b, Megumi Ito 1, Michael Shake 66/1.53, Mr Brightside 189/5.43, Rob Byron 66/1.55, Skelmine Krill 135, Steve Carroll 195/5.63, StillFX 66/1.57, Yobidaba 30r, Yuri Arcurs 14 (INSET); **TLC (Southern) Ltd**: 205/6.6; **Will Burwell**: 140/4.13, 183/5.24

Cover image: Alamy Images: UpperCut Images

All other images © Pearson Education Ltd / Gareth Boden

Picture Research by: Chrissie Martin

Every effort has been made to trace the copyright holders and we apologise in advance for any unintentional omissions. We would be pleased to insert the appropriate acknowledgement in any subsequent edition of this publication.

Contents

Introduction iv

How this book can help you x

1001 Safe working practices in construction 1

1002 Information, quantities and communicating with others 75

1003 Building methods and construction technology 117

1004 Produce basic woodworking joints 135

1005 Maintain and use carpentry and joinery hand tools 175

1006 Maintain and use carpentry and joinery power tools 203

Index 225

Acknowledgements

Introduction

Welcome to NVQ/SVQ CAA Diploma Level 1 Carpentry & Joinery!

Carpentry & Joinery combines many different practical and visual skills with knowledge of specialised materials and techniques. This book will introduce you to the construction trade and in particular the knowledge and skills needed for interpreting site documents, basic woodworking joints, and the safe storage and use of a selection of hand and power tools.

About this book

This book has been produced to help you build a sound knowledge and understanding of all aspects of the Diploma and NVQ requirements associated with Carpentry & Joinery.

The information in this book covers the information you will need to attain your Level 1 qualification in Carpentry & Joinery. Each chapter of the book relates to a particular unit of the CAA Diploma and provides the information needed to form the required knowledge and understanding of that area. The book is also designed to support those undertaking the NVQ at Level 1.

This book has been written based on a concept used with Carillion Training Centres for many years. The concept is about providing learners with the necessary information they need to support their studies and at the same time ensuring it is presented in a style which is both manageable and relevant.

This book will also be a useful reference tool for you in your professional life once you have gained your qualifications and are a practising carpenter or joiner.

This introduction will introduce the construction industry and the qualifications you can find in it, alongside the qualifications available.

About the construction industry

Construction means creating buildings and services. These might be houses, hospitals, schools, offices, roads, bridges, museums, prisons, train stations, airports, monuments – and anything else you can think of that needs designing and building! What about an Olympic stadium? The 2012 London games will bring a wealth of construction opportunity to the UK, so it is an exciting time to be getting involved.

In the UK, 2.2 million people work in the construction industry – more than any other – and it is constantly expanding and developing. There are more choices and opportunities than ever before. Your career doesn't have to end in the UK either – what about taking the skills and experience you are developing abroad? Construction is a career you can take with you wherever you go. There's always going to be something that needs building!

The construction industry is made up of countless companies and businesses that all provide different services and materials. An easy way to divide these companies into categories is according to their size.

- A small company is defined as having between 1 and 49 members of staff.
- A medium company consists of between 50 and 249 members of staff.
- A large company has 250 or more people working for it.

A business might even consist of only one member of staff (a sole trader).

Different types of construction work

There are four main types of construction work:

- New work – this refers to a building that is about to be or has just been built.
- Maintenance work – this is when an existing building is kept up to an acceptable standard by fixing anything that is damaged so that it does not fall into disrepair.
- Refurbishment/renovation work – this generally refers to an existing building that has fallen into a state of disrepair and is then brought up to standard by repair. It also refers to an existing building that is to be used for a different purpose, for example changing an old bank into a pub.
- Restoration work – this refers to an existing building that has fallen into a state of disrepair and is then brought back to its original condition or use.

These four types of work can fall into one of two categories depending upon who is paying for the work:

- Public – the government pays for the work, as is the case with most schools and hospitals etc.
- Private – work is paid for by a private client and can range from extensions on existing houses to new houses or buildings.

Jobs and careers

Jobs and careers in the construction industry fall mainly into one of four categories:

- **building** – the physical construction (making) of a structure. It also involves the maintenance, restoration and refurbishment of structures.

- **civil engineering** – the construction and maintenance of work, such as roads, railways, bridges etc.
- **electrical engineering** – the installation and maintenance of electrical systems and devices such as lights, power sockets and electrical appliances etc.
- **mechanical engineering** – the installation and maintenance of things such as heating, ventilation and lifts

The category that is the most relevant to your course is building.

What is a building?

There are of course lots of very different types of building, but the main types are:

- **residential** – houses, flats etc.
- **commercial** – shops, supermarkets etc.
- **industrial** – warehouses, factories etc.

These types of building can be further broken down by the height or number of storeys that they have (one storey being the level from floor to ceiling):

- **low rise** – a building with one to three storeys
- **medium rise** – a building with four to seven storeys
- **high rise** – a building with seven storeys or more.

Buildings can also be categorised according to the number of other buildings they are attached to:

- **detached** – a building that stands alone and is not connected to any other building
- **semi-detached** – a building that is joined to one other building and shares a dividing wall, called a party wall
- **terraced** – a row of three or more buildings that are joined together, of which the inner buildings share two party walls.

Building requirements

Every building must meet the minimum requirements of the *HAVE*, which were first introduced in 1961 and then updated in 1985. The purpose of building regulations is to ensure that safe and healthy buildings are constructed for the public and that **conservation** (the preservation of the environment and the wildlife) is taken into account when they are being constructed. Building regulations enforce a minimum standard of building work and ensure that the materials used are of a good standard and fit for purpose.

What makes a good building?

When a building is designed, there are certain things that need to be taken into consideration, such as:

- security
- warmth
- safety
- light
- privacy
- ventilation

A well-designed building will meet the minimum standards for all of the considerations above and will also be built in line with building regulations.

Qualifications for the construction industry

There are many ways of entering the construction industry, but the most common method is as an apprentice.

Apprenticeships

You can become an apprentice by being employed:

- directly by a construction company who will send you to college
- by a training provider, such as Carillion, which combines construction training with practical work experience.

Construction Skills is the national training organisation for construction in the UK and is responsible for setting training standards.

The framework of an apprenticeship is based around an NVQ (or SVQ in Scotland). These qualifications are developed and approved by industry experts and will measure your practical skills and job knowledge on-site.

You will also need:

- a technical certificate
- the Construction Skills health and safety test
- the appropriate level of Functional skills assessment
- an Employees Rights and Responsibilities briefing.

You will also need to achieve the right qualifications to get on a construction site, including qualifying for the CSCS card scheme.

CAA Diploma

The Construction Awards Alliance (CAA) Diploma was launched on 1 August 2008 to replace Construction Awards. They aim to make you:

- more skilled and knowledgeable
- more confident with moving across projects, contracts and employers

The CAA Diploma is a common testing strategy with knowledge tests for each unit, a practical assignment and the GOLA (Global Online Assessment) test.

The CAA Diploma meets the requirements of the new Qualifications and Credit Framework (QCF) which bases a qualification on the number of credits gained (with ten learning hours making up one credit):

* Award (1 to 12 credits)
* Certificate (13 to 36 credits)
* Diploma (37+ credits)

As part of the CAA Diploma you will gain the skills needed for the NVQ as well as the Functional skills knowledge you will need to complete your qualification.

National Vocational Qualifications (NVQs)

NVQs are available to anyone, with no restrictions on age, length or type of training, although learners below a certain age can only perform certain tasks. There are different levels of NVQ (for example 1, 2, 3), which in turn are broken down into units of competence. NVQs are not like traditional examinations in which someone sits an exam paper. An NVQ is a 'doing' qualification, which means it lets the industry know that you have the knowledge, skills and ability to actually 'do' something.

NVQs are made up of both mandatory and optional units and the number of units that you need to complete for an NVQ depends on the level and the occupation.

NVQs are assessed in the workplace, and several types of evidence are used:

* Witness testimony provided by individuals who have first-hand knowledge of your work and performance relating to the NVQ.
* Your performance can be observed a number of times in the workplace.
* Historical evidence means that you can use evidence from past achievements or experience, if it is directly related to the NVQ.
* Assignments or projects can be used to assess your knowledge and understanding.
* Photographic evidence showing you performing various tasks in the workplace can be used, providing it is authenticated by your supervisor.

Functional Skills

Throughout this book you will find references to Functional Skills.

Functional skills are processes of representing, analysing and interpreting information. They are the skills needed to work independently in everyday life and are transferable to any given context. We will focus on the mathematics and English skills specifically in a construction context, so that you can identify and practise them in working through the units of the book. The references are headed **FM** for mathematics and **FE** for English. You will also use speaking and listening skills in your learning, which will support you through the programme. If you have any questions on how the skills fit into your learning please speak to your tutor(s).

Introduction

How this book can help you

This book has been fully illustrated with artworks and photographs. These will help to give you more information about a concept or a procedure, as well as helping you to follow a step-by-step procedure or identify a tool or material.

This book also contains a number of different features to help your learning and development.

Functional skills

Functional skills are the skills needed to work independently in everyday life. The references are headed FM for mathematics and FE for English

Key term

These are new or difficult words. They are picked out in **bold** in the text and then defined in the margin

Safety tip

This feature gives you guidance for working safely on the tasks in this book

Did you know?

This feature gives you interesting facts about the building trade

Remember

This highlights key facts or concepts, sometimes from earlier in the text, to remind you of important things you will need to think about

Find out

These are short activities and research opportunities, designed to help you gain further information about, and understanding of, a topic area

Working Life

This feature gives you a chance to read about and debate a real life work scenario or problem. Why has the situation occurred? What would you do?

FAQ

These are frequently asked questions appearing at the end of each unit to answer your questions with informative answers from the experts.

Check it out

A series of questions at the end of each unit to check your understanding. Some of these questions may support the collecting of evidence for the NVQ.

Getting ready for assessment

This feature provides guidance for preparing for the practical assessment. It will give you advice on using the theory you have learnt in a practical way.

CHECK YOUR KNOWLEDGE

This is a series of multiple-choice questions in the style of the GOLA end of unit tests at the end of each unit.

Unit 1001

Safe working practices in construction

Health and safety is a vital part of all construction work. All work should be completed in a way that is safe not only for the individual worker, but also for the other workers on the site, people near by and the final users of the building.

Every year in the construction industry over 100 people are killed and many more are seriously injured as a result of the work that they do. There are thousands more who suffer from work-related health problems such as dermatitis, asbestosis, industrial asthma, vibration white finger (see pages 31–32) or deafness. Therefore, learning as much as you can about health and safety is very important.

This unit also supports NVQ Unit VR1 Conform to General Workplace Safety and VR 03 Move and Handle Resources.

This unit contains material that supports TAP Unit 1 Erect and Dismantle Working Platforms. It also contains material that supports the delivery of the five generic units.

This unit will cover the following learning outcomes:

- Health and safety regulations – roles and responsibilities
- Accident, first aid and emergency procedures and reporting
- Hazards on construction sites
- Health and hygiene
- Safe handling of materials and equipment
- Basic working platforms
- Working with electricity
- Use of appropriate personal protective equipment (PPE)
- Fire and emergency procedures
- Safety signs and notices

Key terms

Legislation – a law or set of laws passed by Parliament, often called an Act

Hazardous – something or a situation that is dangerous or unsafe

Employer – the person or company you work for

Employee – the worker

Proactive – acting in advance, before something happens (such as an accident)

Reactive – acting after something happens, in response to it

Functional skills

When reading and understanding the text in this unit, you are practising several functional skills.

FE 1.2.1 - Identifying how the main points and ideas are organised in different texts.

FE 1.2.2 – Understanding different texts in detail.

FE 1.2.3 – Read different texts and take appropriate action, e.g. respond to advice/instructions.

If there are any words or phrases you do not understand, use a dictionary, look them up using the internet or discuss with your tutor.

K1. Health and safety regulations

While at work, whatever your location or the type of work you are doing, there is important **legislation** you must comply with. Health and safety legislation is there not just to protect you – it also states what you must and must not do to ensure that no workers are placed in a situation **hazardous** to themselves or others.

There are hundreds of Acts covering all manner of work from hairdressing to construction. Each Act states the duties of the **employer** and **employee** – and you should be aware of both. If an employer or employee does something they shouldn't – or doesn't do something they should – they can end up in court and be fined or even imprisoned.

Approved code of practice, guidance notes and safety policies

As well as Acts, there are two sorts of codes of practice and guidance notes: those produced by the Health and Safety Executive (HSE; see page 5), and those created by companies themselves. Most large construction companies – and many smaller ones – have their own guidance notes, which go further than health and safety law. For example, the law states that everyone must wear safety boots in a hazardous area, but a company's code may state that everyone must wear safety boots at all times. This is called taking a **proactive** approach, rather than a **reactive** one.

Most companies have some form of safety policy outlining the company's commitment and stating what they plan to do to ensure that all work is carried out as safely as possible. As an employee, you should make sure you understand the company's safety policy as well as its codes of practice. If you don't follow company policy you may not be prosecuted in court, but you could still be disciplined by the company or even fired.

Health and safety legislation you need to be aware of

There are some 20 pieces of legislation you will need to be aware of, each of which sets out requirements for employers and often for employees.

Example

The Health and Safety at Work Act 1974 states that an employer must 'so far as is reasonably practicable' ensure that a safe place of work is provided. Yet employers are not expected to do everything they can to protect their staff from lightning strikes, as there is only a 1 in 800,000 chance of this occurring – this would not be reasonable!

We will now look at the regulations that will affect you most.

The Health and Safety at Work Act 1974 (HASAW)

HASAW applies to all types and places of work and to employers, employees, self-employed people, **subcontractors** and even **suppliers**. The Act is there to protect not only the people at work but also the general public, who may be affected in some way by the work that has been or will be carried out.

Key terms

Subcontractor – workers who have been hired by the main contractor to carry out works, usually specialist works, e.g. a general builder may hire a plumber as a subcontractor as none of their staff can do plumbing work

Supplier – a company that supplies goods, materials or services

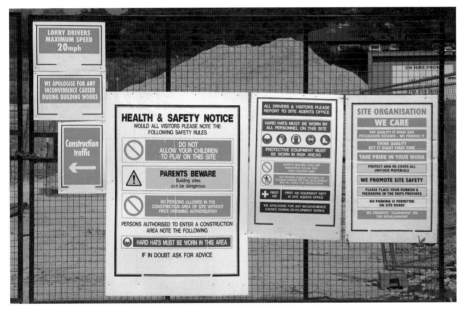

Figure 1.1 Legislation is there to protect employees and the public alike

Did you know?

One phrase that often comes up in the legislation is 'so far as is reasonably practicable'. This means that health and safety must be adhered to at all times, but must take a common-sense, practical approach

The main objectives of the Health and Safety at Work Act are to:

- ensure the health, safety and welfare of all persons at work
- protect the general public from all work activities
- control the use, handling, storage and transportation of explosives and highly **flammable** substances
- control the release of noxious or offensive substances into the atmosphere.

To ensure that these objectives are met there are duties for all employers, employees and suppliers.

Key term

Flammable – something that is easily lit and burns rapidly

Key terms

Access – entrance, a way in

Egress – exit, a way out

PPE – personal protective equipment, such as gloves, a safety harness or goggles

The employer's duties

Employers must:

- provide safe **access** and **egress** to and within the work area
- provide a safe place to work
- provide and maintain plant and machinery that is safe and without risks to health
- provide information, instruction, training and supervision to ensure the health and safety at work of all employees
- ensure safety and the absence of risks to health in connection with the handling, storage and transportation of articles and substances
- have a written safety policy that must be revised and updated regularly, and ensure all employees are aware of it
- involve trade union safety representatives, where appointed, in all matters relating to health and safety
- carry out risk assessments (see pages 24–25) and provide supervision where necessary
- provide and not charge for personal protective equipment (**PPE**).

The employee's duties

Employees must:

- take reasonable care for their own health and safety
- take reasonable care for the health and safety of anyone who may be affected by their acts or **omissions**
- co-operate with their employer or any other person to ensure the legal **obligations** are met
- not misuse or interfere with anything provided for their health and safety
- report hazards and accidents (see page 15)
- use any equipment and safeguards provided by their employer.

Key terms

Omission – something that has not been done or has been missed out

Obligation – something you have a duty or a responsibility to do

Employers can't charge their employees for anything that has been done or provided for them to ensure that legal requirements on health and safety are met. Self-employed people and subcontractors have the same duties as employees. If they have employees of their own, they must also obey the duties set down for employers.

The supplier's duties

Persons designing, manufacturing, importing or supplying articles or substances for use at work must ensure that:

- articles are designed and constructed so that they will be safe and without risk to health while they are being used or constructed

- substances will be safe and without risk to health at all times when being used, handled, transported and stored

- tests on articles and substances are carried out as necessary

- adequate information is provided about the use, handling, transporting and storage of articles or substances.

Health and Safety Executive (HSE)

HASAW, like most of the other Acts mentioned, is enforced by the HSE.

The HSE is the government body responsible for the encouragement, regulation and enforcement of health, safety and welfare in the workplace in the UK. It also has responsibility for research into occupational risks in England, Wales and Scotland. In Northern Ireland the responsibility lies with the Health and Safety Executive for Northern Ireland.

The HSE's duties are to:

- assist and encourage anyone who has any dealings with the objectives of HASAW

- produce and encourage research, publication, training and information on health and safety at work

- ensure that employers, employees, suppliers and other people are provided with an information and advisory service, and are kept informed and advised on any health and safety matters

- propose regulations

- enforce HASAW.

To aid in these duties the HSE has several resources, including a laboratory used for, among other things, research, development and **forensic investigation** into the causes of accidents. The enforcement of HASAW is usually delegated to local government bodies such as county or district councils.

Local government bodies can be **enforcing authorities** for several workplaces, including offices, shops, retail and wholesale distribution, hotel and catering establishments, petrol filling stations, residential care homes and the leisure industry.

An enforcing authority may appoint **inspectors**, who, under the authority, have the power to:

- enter any premises which she or he has reason to believe it is necessary to enter so enforce the Act, at any reasonable time, or in a dangerous situation

- bring a police constable if there is reasonable cause to fear any serious obstruction in carrying out their duty

Key terms

Forensic investigation – a branch of science that looks at how and why things happen

Enforcing authorities – an organisation or people who have the authority to enforce certain laws or Acts, as well as providing guidance or advice

Inspector – someone who is appointed or employed to inspect/examine something in order to judge its quality or compliance with any laws

- bring any other person authorised by the enforcing authority; and any equipment or materials required
- examine and investigate any circumstance that is necessary for the purpose of enforcing the Act
- give orders that the premises, any part of them or anything therein, shall be left undisturbed for as long as needed for the purpose of any examination or investigation
- take measurements, photographs and make any recordings considered necessary for the purpose of examination or investigation
- take samples of any articles or substances found and of the atmosphere in or in the vicinity of the premises
- have an article or substance which appears to be a danger to health or safety, dismantled, tested or even destroyed if necessary
- take possession of such an article and detain it for as long as necessary in order to examine it and ensure that it is not tampered with and that it is available for use as evidence in any **prosecution**
- interview any person believed to have information, ask any questions the inspector thinks fit to ask and ensure all statements are signed as a declaration of the truth of the answers
- require the production of, inspect and take copies of, any entry in any book or document which it is necessary for the purposes of any examination or investigation
- any other power which is necessary to enforce the Act.

Contacting the HSE

Employers, self-employed people or those in control of work premises have legal duties to record and report to the HSE some work-related accidents. Incidents that must be reported are:

- **death** – where someone is killed as a result of an accident related to work; this includes deaths resulted from physical violence
- **major injury** – this includes fractures, amputations, loss of sight and loss of consciousness
- **dangerous occurrence** – this is an event that may not have caused injury, but clearly could have done so. For example, some kinds of fire or explosion, collapse of buildings or scaffolding
- **over three day injury** – where someone suffers an injury at work that results in them being away from work or unable to perform their full duties for more than three consecutive days
- **disease** – where a doctor notifies the person that they are suffering from some work-related disease.

Key term

Prosecution – accusing someone of committing a crime, which usually results in the accused being taken to court and, if found guilty, being punished

Reporting of Injuries, Diseases and Dangerous Occurrences Regulations 1995 (RIDDOR)

Under RIDDOR, employers have a duty to report accidents, diseases or dangerous occurrences. The HSE uses this information to identify where and how risk arises and to investigate serious accidents.

Control of Substances Hazardous to Health Regulations 2002 (COSHH)

These regulations state how employees and employers should work with, handle, store, transport and dispose of potentially hazardous substances (substances that might adversely affect your health) including:

- substances used directly in work activities (e.g. adhesives or paints)
- substances generated during work activities (e.g. dust from sanding wood)
- naturally occurring substances (e.g. sand dust)
- biological agents (e.g. germs such as bacteria).

These substances can be found in nearly all work environments.

The Control of Noise at Work Regulations 2005

At some point in your career in construction, you are likely to work in a noisy working environment. These regulations help protect you against the consequences of being exposed to high levels of noise. High levels of noise can lead to permanent hearing damage. These regulations state that the employer must:

- assess the risks to the employee from noise at work
- take action to reduce the noise exposure that produces these risks
- provide employees with hearing protection or, if this is impossible, reduce the risk by other methods
- make sure the legal limits on noise exposure are not exceeded
- provide employees with information, instruction and training
- carry out **health surveillance** where there is a risk to health.

> **Remember**
>
> All hazardous substances are covered by COSHH regulations, except asbestos and lead paint, which have their own regulations

Figure 1.2 Noise at work

> **Key term**
>
> **Health surveillance** – where a company will assess the risks of tasks that are to be done to see if these tasks will create risks to health

The Electricity at Work Regulations 1989

These regulations cover any work involving the use of electricity or electrical equipment. An employer has the duty to ensure that the electrical systems their employees come into contact with are safe and regularly maintained. They must also have done everything the law states to reduce the risk of their employees coming into contact with live electrical currents.

Construction (Design and Management) Regulations 2007

The Construction (Design and Management) Regulations 2007, often referred to as CDM, are important regulations in the construction industry. They were introduced by the HSE's Construction Division. The regulations deal mainly with the construction industry and aim to improve safety.

The duties for employers under the regulations are to:

- plan, manage and monitor own work and that of workers
- check competence of all their appointees and workers
- train their employees
- provide information to their workers
- comply with the specific requirements in Part 4 of the Regulations, which deals with lighting, excavations, traffic routes, etc.
- ensure there are adequate welfare facilities for their workers.

The duties for employees are to:

- check their own competence
- co-operate with others and co-ordinate work so as to ensure the health and safety of construction workers and others who may be affected by the work
- report obvious risks.

The CDM also requires certain duties from the clients (with the exception of domestic clients). These duties are to:

- check competence and resources of all appointees
- ensure there are suitable management arrangements for the project welfare facilities
- allow sufficient time and resources for all stages
- provide pre-construction information to designers and contractors.

There is a general expectation by the HSE that all parties involved in a project will co-operate and co-ordinate with each other.

Did you know?

On large projects, a person is appointed as the CDM co-ordinator. This person has overall responsibility for compliance with CDM

Provision and Use of Work Equipment Regulations 1998 (PUWER)

These regulations cover all new or existing work equipment – leased, hired or second-hand. They apply in most working environments where the HASAW applies, including all industrial, offshore and service operations.

PUWER covers starting, stopping, regular use, transport, repair, modification, servicing and cleaning.

The general duties of the Act require equipment to be:

- suitable for its intended purpose and only to be used in suitable conditions
- maintained in an efficient state and maintenance records kept
- used, repaired and maintained only by a suitably trained person, when that equipment poses a particular risk
- able to be isolated from all its sources of energy
- constructed or adapted to ensure that maintenance can be carried out without risks to health and safety
- fitted with warnings or warning devices as appropriate.

In addition, the Act requires:

- all those who use, supervise or manage work equipment to be suitably trained
- access to any dangerous parts of the machinery to be prevented or controlled
- injury to be prevented from any work equipment that may have a very high or low temperature
- suitable controls to be provided for starting and stopping the work equipment
- suitable emergency stopping systems and braking systems to be fitted to ensure the work equipment is brought to a safe condition as soon as reasonably practicable
- suitable and sufficient lighting to be provided for operating the work equipment.

Manual Handling Operations Regulations 1992

These regulations cover all work activities in which a person rather than a machine does the lifting. The regulations state that, wherever possible, manual handling should be avoided, but where this is unavoidable, a risk assessment should be done.

In a risk assessment, there are four considerations:

- **load** – is it heavy, sharp-edged, difficult to hold?

Did you know?

'Work equipment' includes any machinery, appliance, apparatus or tool, and any assembly of components that are used in non-domestic premises. Dumper trucks, circular saws, ladders, overhead projectors and chisels would all be included, but substances, private cars and structural items fall outside this definition

- **individual** – is the individual small, pregnant, in need of training?
- **task** – does the task require holding goods away from the body, or repetitive twisting?
- **environment** – is the floor uneven, are there stairs, is it raining?

After the assessment, the situation must be monitored constantly and updated or changed if necessary.

Personal Protective Equipment at Work Regulations 1992 (PPER)

These regulations cover all types of PPE, from gloves to breathing apparatus. After doing a risk assessment and once the potential hazards are known, suitable types of PPE can be selected. PPE should be checked prior to issue by a trained and competent person and in line with the manufacturer's instructions. Where required, the employer must provide PPE free of charge along with a suitable and secure place to store it.

The employer must ensure that the employee knows:

- the risks the PPE will avoid or reduce
- its purpose and use
- how to maintain and look after it
- its limitations.

The employee must:

- ensure that they are trained in the use of the PPE prior to use
- use it in line with the employer's instructions
- return it to storage after use
- take care of it, and report any loss or defect to their employer.

> **Remember**
>
> PPE must only be used as a last line of defence

Work at Height Regulations 2005

Construction workers often work high off the ground, on scaffolding, ladders or roofs. These regulations make sure that employers do all that they can to reduce the risk of injury or death from working at height.

The employer has a duty to:

- avoid work at height where possible
- use any equipment or safeguards that will prevent falls
- use equipment and any other methods that will minimise the distance and consequences of a fall.

As an employee, you must follow any training given to you, report any hazards to your supervisor and use any safety equipment provided to you.

Other Acts to be aware of

You should also be aware of the following pieces of legislation:

- The Fire Precautions (Workplace) Regulations 1997
- The Fire Precautions Act 1991
- The Highly Flammable Liquids and Liquid Petroleum Gases Regulations 1972
- The Lifting Operations and Lifting Equipment Regulations 1998
- The Construction (Health, Safety and Welfare) Regulations 1996
- The Environmental Protection Act 1990
- The Confined Spaces Regulations 1997
- The Working Time Regulations 1998
- The Health and Safety (First Aid) Regulations 1981.

You can find out more at the library or online.

Sources of health and safety information

Health and safety is a large and varied subject that changes regularly. The introduction of new regulations or updates to current legislation means it's often hard to remember or keep up to date. Your tutor will be able to give you information on current legislation.

Your employer should keep you updated on any changes to legislation that will affect you. You can also access other sources of information to keep you informed.

Health and Safety Executive

The HSE has a wide range of information ranging from the actual legislation documents to helpful guides to working safely. Videos, leaflets and documents are available to download free from its website. Specific sections of the website are dedicated to different industries ranging from agriculture to hairdressing. The specific construction website address is www.hse.gov.uk/construction.

ConstructionSkills

ConstructionSkills mainly offers advice on qualifications in construction. However, it also has advice on health and safety matters and on sitting the CSCS (Construction Skills Certification Scheme) health and safety test as well as providing a way of booking the test. The website address is www.cskills.org.

Find out

Look into the other regulations listed here via the HSE website (www.hse.gov.uk)

Remember

Legislation can change or be updated. New legislation can be created as well – this could even supersede all pieces of legislation

Did you know?

The HSE now also includes the Health and Safety Commission (HSC), which was merged with it in 2008

Royal Society for the Prevention of Accidents (RoSPA)

RoSPA provides information, advice, resources and training. RoSPA is actively involved in the promotion of safety and the prevention of accidents in all areas of life – at work, in the home, on the roads, in schools, at leisure and on (or near) water. The website address is www.rospa.com.

Royal Society for the Promotion of Health (RSPH)

The Royal Society for the Promotion of Health aims to promote and protect health and well-being. It uses **advocacy**, mediation, knowledge and practice to advise on policy development. It also provides education and training services, encourages research, communicates information and provides certification to products, training centres and processes.

RSPH's main focus is people working in healthcare, for example doctors. It publishes two journals:

- *Public health*
- *Perspectives on Public Health*.

Site inductions

Key term

Advocacy – actively supporting or arguing for

Figure 1.3 A site induction talk

Key term

Induction – a formal introduction you will receive when you start any new job, where you will be shown around, shown where the toilets and canteen etc. are, and told what to do if there is a fire

Site **induction** is the process that an individual undergoes in order to accelerate their awareness of the potential health and safety hazards and risks they may face in their working environment. Site induction doesn't include job-related skills training.

Different site inductions will cover different topics, depending on the work that is being carried out. The basic information that should be included in inductions will cover:

- the scope of operations carried out at the site, project, etc.
- the activities that have health and safety hazards and risks
- the control measures in place
- emergency arrangements
- local organisation and management structure
- consultation procedures, including 'Don't walk by. Take action now' and Safety Action Groups
- the resource for health and safety advice
- the welfare arrangements
- a zero-tolerance approach to health and safety risks at work
- the process for reporting near misses (see page 17).

Inductions are also vital for informing all people working on the site of amenities, restricted areas, dress code (PPE) and evacuation procedures. Inductions must be carried out by a competent person. Records of all inductions must be kept to ensure that all workers have received an induction. Some sites will even hand out cards to those who have been inducted and people without cards will not be admitted to the site.

Visitors to the site who may not be actually doing any work should still receive an induction of sorts as they also need to be aware of amenities, restricted areas and procedures.

Toolbox talks

Toolbox talks are used by management, supervisors and employees to deliver basic training and/or to inform all workers of any updates to policy, hazardous activities/areas or any other information.

Toolbox talk topics should be relevant to the people they are being delivered to (there is no point delivering a talk on plumbing systems to bricklayers unless it directly affects them!). The topics can vary from being informative, such as letting everyone know a reclassification of a PPE area, to basic training on the use of a certain tool.

They should be delivered by a competent person and a record of all attendees should be kept.

Remember

A site induction must take place *before* you start work on that site

Did you know?

Toolbox talks don't need to be formal meetings but can be held in a canteen at break time. However, a list of all attendees must be kept to ensure that everyone who needs to receive the talk does so

Working Life

Alex and Molly have been asked to attend a toolbox talk on scaffolding safety. They had both attended a toolbox talk on the same subject just over a week ago. Molly thinks they should attend this talk too as it could be important. However, Alex thinks it's a mistake and that it will be the same as the last one. Molly agrees that they are very busy and that if they don't attend they can get the job they are doing finished on time; but she is concerned about missing the talk.

- Why could Alex and Molly be asked to attend if they had a similarly titled toolbox talk recently?
- What could the outcome be if they do/don't attend?
- What things could be discussed in a toolbox talk on scaffolding?

Construction Skills Certification Scheme (CSCS)

Figure 1.4 A CSCS card

The Construction Skills Certification Scheme was introduced to help improve the quality of work and to reduce accidents. It requires all workers to obtain a CSCS card before they are allowed to carry out work on a building site. There are various levels of card, which indicate your competence and skills background. This ensures that only skilled and safe tradespeople can carry out the required work on site.

To get a CSCS card all applicants must sit a health and safety test. The aim of the test is to examine knowledge across a wide range of topics to improve safety and productivity on site. You usually take the test as a PC-based touch screen test, either at a mobile testing unit or an accredited test centre. The type of card you apply for will determine the level of test that you need to take.

As a trainee, once you pass the health and safety test you will qualify for a trainee card. Once you have achieved a Level 2 qualification you can then upgrade your card to an experienced worker card. Achieving a Level 3 qualification allows you to apply for a gold card. People who make regular visits to site can apply for a visitor card.

K2. Accident, first aid and emergency procedures and reporting

Major types of emergency

There are several types of major emergency that could occur on site. These include not only accidents but also:

- fires
- security alerts
- bomb scares.

At your site induction, it should be made perfectly clear to you what you should do in an emergency. You should also be aware of any sirens or warning noises that accompany each and every type of emergency such as bomb scares or fire alarms. Some sites may have different variations on sirens or emergency procedures, so it is vital that you pay attention and listen to all the instructions. If you are unsure always ask.

The key legislation that controls the reporting of accidents is RIDDOR.

Health and welfare in the construction industry

Jobs in the construction industry have one of the highest injury and accident rates. As a worker you will be at constant risk unless you adopt a good health and safety attitude. By following the rules and regulations set out to protect you and by taking reasonable care of yourself and others, you will become a safe worker and thus reduce the chance of any injuries or accidents.

Accidents

We often hear the saying 'accidents will happen', but when working in the construction industry, we should not accept that accidents just happen sometimes. When we think of an accident, we quite often think about it as being no one's fault and something that could not have been avoided. The truth is that most accidents are caused by human error, which means someone has done something they shouldn't have done or, just as importantly, not done something they should have done.

Accidents often happen when someone:

- is hurrying
- not paying enough attention to what they are doing
- has not received the correct training.

Reporting accidents

When an accident occurs, there are certain things that must be done. All accidents need to be reported and recorded in the accident book. The injured person

Remember

Health and safety laws are there to protect you and other people. If you take shortcuts or ignore the rules, you are placing yourself and others at serious risk

Figure 1.5 Accidents can happen if your work area is untidy

must report to a trained first aider to receive treatment. Serious accidents must be reported under RIDDOR.

Accidents and emergencies must be reported to the relevant authorised persons. These can be:

- **first aiders** – all accidents need to be reported to a first aider. If you are unsure who the first aiders are or have no direct way of contacting them, you must report it to your supervisor
- **supervisors** – you must inform your supervisor of any accident as it is vital that they can act immediately to inform the relevant first aider or their manager, and stop the work if necessary to prevent any further accidents
- **safety officers** – your supervisor or the site manager will alert the safety officer who will assess the area to check if it is safe, investigate what may have caused the accident and prepare reports for the HSE (if needed)
- **HSE** – if death or major injury occurs to a member of staff or a member of the public is killed or taken to hospital the accident must be reported to the HSE immediately, and followed up by a written report within ten days. The written report is made on form F2508. If an employee suffers an 'over-three-day' injury it must be reported on the F2508 form within ten days
- **managers** – managers should be informed by either the supervisor or safety officer as may need to report to head office. They may also be the one tasked with contacting the HSE
- **emergency services** – the emergency services should be called as soon as possible. Usually the first aiders will call the ambulance and the supervisors will call the fire brigade, but if in doubt you should also call.

Under RIDDOR your employer must report to the HSE any accident that results in:

- death
- major injury
- an injury that means the injured person is not at work for more than three consecutive days
- disease.

Diseases that can be caused in the workplace include:

- certain poisonings
- some skin diseases – such as occupational dermatitis, skin cancer, chrome ulcer, oil folliculitis/acne
- lung diseases – including occupational asthma, farmer's lung, pneumoconiosis, asbestosis, mesothelioma

Remember

An accident that falls under RIDDOR should be reported by the safety officer or site manager. It can be reported to the HSE on phone (0845 3009923) or via the RIDDOR website (www. riddor.gov.uk)

Safety tip

The emergency services would rather be called twice than not at all

Unit 1001 Safe working practices in construction

- infections – such as leptospirosis (see page 31), hepatitis, tuberculosis, anthrax, legionellosis and tetanus
- other conditions – such as occupational cancer, certain musculoskeletal disorders, decompression illness and hand–arm vibration syndrome (see page 31).

The nature and seriousness of the accident will determine who it needs to be reported to. There are several types of documentation used to record accidents and emergencies.

The accident book

The accident book is completed by the person who had the accident or, if this is not possible, someone who is representing the injured person. The accident book will ask for some basic details about the accident, including:

- who was involved
- what happened
- where it happened
- the day and time of the accident
- any witnesses to the accident
- the address of the injured person
- what PPE was being worn
- what first aid treatment was given.

Major and minor accidents

If an accident happens, you or the person it happened to may be lucky and will not be injured. More often, an accident will result in an injury which may be minor (e.g. a cut or a bruise) or possibly major (e.g. loss of a limb). Accidents can also be fatal. The most common causes of fatal accidents in the construction industry are:

- falling from scaffolding
- being hit by falling objects and materials
- falling through fragile roofs
- being hit by forklifts or lorries
- cuts
- infections
- burns
- electrocution.

Near misses

As well as reporting accidents, 'near misses' must also be reported. A 'near miss' is when an accident nearly happened but did not actually occur. Reporting near misses might identify a

Find out

Visit the NHS Choices website (www.nhs.uk) to find out more about these diseases, and others that can be caused in the workplace

Safety tip

Near misses must be recorded because they are often the accidents of the future

problem and can prevent accidents from happening in the future. This allows a company to be proactive rather than reactive.

Work-related injuries in the construction industry

Construction has the largest number of fatal injuries of all the main industry groups. In 2007–2008 there were 72 fatal injuries. This gave a rate of 3.4 people injured per 100 000 workers. The rate of fatal injuries in construction over the past decade has shown a downward trend, however, the rate has shown little change in the most recent years.

- From 1999–2000 to 2006–2007 the rate of reported major injuries in construction fell. It is unclear whether the rise in 2007–2008 means an end to this trend. Despite this falling trend, the rate of major injury in construction is the highest of any main industry group (599.2 per 100 000 employees in 2007–2008).

- A higher proportion of the reported injuries in construction were caused by falls from height, falling objects and contact with moving machinery.

- The THOR-GP surveillance scheme data (2006–2007) indicates a higher rate of work-related illness in construction than across all industries. The rate of self-reported work-related ill health in construction is similar to other industries.

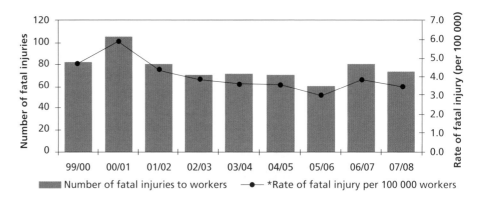

Figure 1.6 Number and rate of fatal injury to workers, 1999–2000 to 2007–2008

The cost of accidents

As well as the tragedy, pain and suffering that accidents cause, they can also have a negative financial and business impact.

Small accidents will affect profits as sick pay may need to be paid. Production may also slow down or stop if the injured person is a specialist. Replacement or temporary workers may need to be used to keep the job going. This can cost small companies with a handful of employees hundreds of pounds for every day

Remember

Clients don't want to hire companies that are not deemed safe

that an injured person can't work. Larger companies with many employees may have several people off work at once which can cost thousands of pounds per day.

More serious accidents will see the financial loss rise as the injured person will be off work for longer. This can cause jobs to fall seriously behind and, in extreme cases, may even cause the contractor to lose the job and possibly have to close the site.

Companies that have a lot of accidents will have a poor company image for health and safety. They will also find it increasingly difficult to gain future contracts. Unsafe companies with lots of accidents will also see injured people claiming against their insurance, which will see their premiums rise. This will eventually make them uninsurable, meaning they will not get any work.

First aid

In the unfortunate event of an accident on site, first aid may have to be administered. If there are more than five people on a site, then a qualified first aider must be present at all times. On large building sites there must be several first aiders. During your site induction you will be made aware of who the first aiders are and where the first aid points are situated. A first aid point must have the relevant first aid equipment to deal with the types of injury that are likely to occur. However, first aid is only the first step and, in the case of major injuries, the emergency services should be called.

A good first aid box should have plasters, bandages, antiseptic wipes, latex gloves, eye patches, slings, wound dressings and safety pins. Other equipment such as eye wash stations, must also be available if the work being carried out requires it.

> **Remember**
>
> Health and safety is everyone's duty. If you receive first aid treatment and notice that there are only two plasters left, you should report it to your line manager

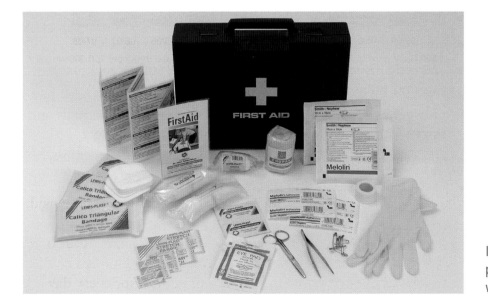

Figure 1.7 A first aid box provides the supplies to deal with minor injuries

Unit 1001 Safe working practices in construction

Actions for an unsafe area

On discovering an accident the first thing to do is to ensure that the victim is in no further danger. This will require you to do tasks such as switching off the electrical supply. Tasks like this must only be done if there is no danger to yourself.

Turning off the electricity is just one possible example. There will be specific safety issues for individual jobs the injured person may have been working on.

You must next contact the first aider. Unless you have been trained in first aid you must not attempt to move the injured person as you may cause more damage. If necessary, the first aider will then call the emergency services.

K3. Hazards on construction sites

A major part of health and safety at work is being able to identify hazards and to help prevent them in the first place, therefore avoiding the risk of injury.

Housekeeping

Housekeeping is the simple term used for cleaning up after yourself to ensure your work area is clear and tidy. Good housekeeping is vital on a construction site, as an unclean work area is dangerous.

To maintain good housekeeping it is important that you:

- work tidily to reduce the risk of you or somebody else getting hurt
- don't overfill skips as this can lead to fire hazards
- ensure fire exits and emergency escape routes are clear
- correctly dispose of food waste as this can attract cockroaches, rats and other vermin
- only get as many nails and screws as you need – loose nails and screws can puncture tyres and even cause injury to feet
- clean and sweep up at the end of each day
- avoid blocking exits and walkways
- be aware while you are working – how might your mess affect you or others?

Storing combustibles and chemicals

It is vital to store combustibles and chemicals on site correctly.

Chemicals

Certain chemicals such as brick cleaner or some types of adhesive are classified as dangerous chemicals. All chemicals should be stored in a locked area to prevent misuse or cross-contamination.

Highly flammable liquids

Liquefied petroleum gas (LPG), petrol, cellulose thinners, methylated spirits, chlorinated rubber paint and white spirit are all highly flammable liquids. These materials require special storage to ensure they do not risk injuring workers.

- Containers should only be kept in a special storeroom built of concrete, brick or some other fireproof material.

- The floor should also be made of concrete and should slope away from the storage area. This is to prevent leaked materials from collecting under the containers.

- The roof should be made from an easily shattered material to minimise the effect of any explosion.

- Doors should be at least 50 mm thick and open outwards.

- Any glass used in the structure should be wired and not less than 6 mm thick.

- The standing area should have a sill surrounding it that is deep enough to contain the contents of the largest container stored.

- Containers should always be stored upright.

- The area should not be heated.

- Electric lights should be safe.

- Light switches should be flameproof and should be on the outside of the store.

- The building should be ventilated at high and low levels and have at least two exits.

- Naked flames and spark-producing materials should be clearly prohibited in the area, including smoking.

- The storeroom should be clearly marked with red and white squares and 'Highly Flammable' signage.

Find out

Check out the storage details for any chemicals that you come across – look at the manufacturer's instructions or the COSHH regulations

Did you know?

'Inflammable' means the same thing as flammable, that is it is easily lit and capable of burning rapidly

Figure 1.8 Storage of highly flammable liquids

In addition to these requirements, there are storage regulations specifically for LPG.

- LPG must be stored in the open and usually in a locked cage.
- It should be stored off the floor and protected from direct sunlight and frost or snow.
- The storage of LPG is governed by the Highly Flammable Liquids and Liquefied Petroleum Gases Regulations. Note that these regulations apply when 50 or more litres are stored, and permission must be obtained from the District Inspector of Factories.

Glass

Glass should be stored vertically in racks. The conditions for glass storage should be:

- clean – storing glass in dirty or dusty locations can cause discoloration
- dry – if moisture is allowed between the sheets of glass it can make them stick together, which may make them difficult to handle and more likely to break.

If only a small number of sheets of glass are to be stored, they can be leant against a stable surface, as shown in Figure 1.9.

Figure 1.9 Storage of glass

Safety tip

Never expose materials such as LPG, white spirit, methylated spirit and turps to a naked flame (including cigarettes) – they are highly flammable

Make sure when working with potentially hazardous materials that you take the appropriate precautions, for example wear gloves and eye protection and work in a ventilated area.

Remember

Always wear appropriate PPE when handling glass

Hazards on the building site

The building industry can be a very dangerous place to work, and there are certain hazards that all workers need to be aware of. The main types of hazard that you will face are:

- falling from height
- tripping
- chemical spills
- burns
- electrocution
- fires.

Falling from height

When working in the construction industry a lot of the work that you do will be at height. The main hazard of working at height is falling. A fall from a scaffold, even if it is at low level, can cause serious injuries such as broken bones. A worker may also suffer permanent damage from the fall. This could leave them wheelchair-bound for life or even kill them.

Tripping

The main cause of tripping is poor housekeeping. Whether working on scaffolding or on ground level, an untidy workplace is an accident waiting to happen. All workplaces should be kept tidy and free of debris. All offcuts should be put either in a wheelbarrow (if you aren't near a skip) or straight into the skip.

Chemical spills

Chemical spillages can range from minor inconvenience to major disaster. Most spillages are small and create minimal or no risk. If the material involved is not hazardous, it simply can be cleaned up by normal operations such as brushing or mopping up the spill. Occasionally, the spill may be on a larger scale and may involve a hazardous material. It is important to know what to do before the spillage happens so that remedial action can be prompt and harmful effects minimised. Of course, when a hazardous substance is being used a COSHH or risk assessment will have been made, and it should include a plan for dealing with a spillage. This in turn should mean that the materials required for dealing with the spillage should be readily available.

Figure 1.10 An untidy work site can present many trip hazards

> **Remember**
>
> Not only will being tidy prevent trip hazards, but it will also prevent costly clean-up operations at the end of the job and will promote a good professional image

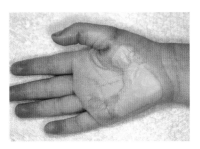

Figure 1.11 Fire, heat, chemicals and electricity can cause burns

Burns

Burns can occur not only from the obvious source of fire and heat but also from materials containing chemicals such as cement or painter's solvents. Even electricity can cause burns. It is vital when working with materials that you are aware of the hazards they may present and take the necessary precautions.

Electricity

Electricity can be very hazardous to work with when proper care is not taken. In your role on a building site you will be working with electricity regularly and should make sure to apply good practice. This has been covered in detail in this unit on page 61.

Fires

The obvious main risk from fires is burns. However during fires, the actual flames are not often the cause of injury or death. Smoke inhalation is a very serious hazard and this is what mainly causes death.

Fires will be covered in greater depth later in this unit (pages 68–70).

Risk assessments

You will have noticed that most of the legislation we have looked at requires risk assessments to be carried out. The Management of Health and Safety at Work Regulations 1999 requires every employer to make suitable and sufficient assessment of:

- the risks to the health and safety of their employees to which the employees are exposed while at work
- the risks to the health and safety of persons not in their employment, arising out of or in connection with their work activities.

It is vital that you know how to **carry out a risk assessment**. Often you may be in a position where you are given direct responsibility for this, and the care and attention you take over it may have a direct impact on the safety of others. You must be aware of the dangers or hazards of any task, and know what can be done to prevent or reduce the risk.

Key term

Carry out a risk assessment – this means measuring the dangers of an activity against the likelihood of accidents taking place

There are five steps in a risk assessment – here we use cutting the grass as an example.

- **Step 1** Identify the hazards

When cutting the grass the main hazards are from the blades or cutting the wire, electrocution and any stones that may be thrown up.

- **Step 2** Identify who will be at risk

The main person at risk is the user, but passers-by may be struck by flying debris.

- **Step 3** Calculate the risk from the hazard against the likelihood of an accident taking place

The risks from the hazard are quite high: the blade or wire can remove a finger, electrocution can kill and the flying debris can blind or even kill. The likelihood of an accident happening is medium: you are unlikely to cut yourself on the blades, but the chance of cutting through the cable is medium, and the chance of hitting a stone high.

- **Step 4** Introduce measures to reduce the risk

Training can reduce the risks of cutting yourself. Training and the use of an **RCD** can reduce the risk of electrocution. Raking the lawn first can reduce the risk of sending up stones.

- **Step 5** Monitor the risk

Constantly changing factors mean any risk assessment may have to be modified or even changed completely. In our example, one such factor could be rain.

Figure 1.12 Even an everyday task like cutting the grass has its own dangers

Key term

RCD – residual current device, a device that will shut the power down on a piece of electrical equipment if it detects a change in the current, thus preventing electrocution

Working Life

Scaffold safety

Ralph and Vijay are working on the second level of some scaffolding clearing debris. Ralph suggests that, to speed up the task, they should throw the debris over the edge of the scaffolding into a skip below. The building Ralph and Vijay are working on is on a main road and the skip is not in a closed off area.

What do you think of Ralph's idea? What are your reasons for this answer?

Method statements

A method statement is a key safety document that takes the information about significant risks from your risk assessment, and combines them with the job specification to produce a practical and safe working method for the workers to follow on site.

Method statements should be specific and relevant to the job in hand and should detail what work is to be done, how the work should be done and what safety precautions need to be taken.

Hazard books

The hazard book is a tool used on some sites to identify hazards within certain tasks. It can also help to produce risk assessments. The book will list tasks and what hazards are associated with those tasks. Different working environments can create different types of hazard so risk assessments must always look at the specific task separately.

K4. Health and hygiene

As well as keeping an eye out for hazards, you must also make sure that you look after yourself and stay healthy. This is a responsibility that lies with both the employer and the employee.

Staying healthy

One of the easiest ways to stay healthy is to wash your hands regularly. By washing your hands you are preventing hazardous substances from entering your body through ingestion (swallowing). You should always wash your hands after going to the toilet and before eating or drinking. Personal hygiene is vital to ensure good health.

Remember that some health problems do not show symptoms straight away and what you do now can affect you much later in life.

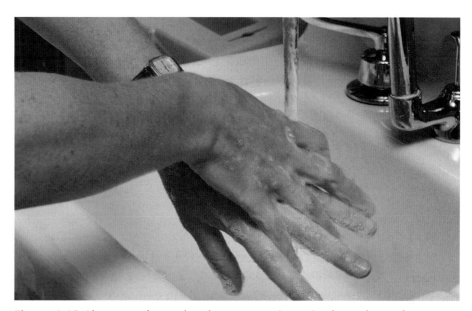

Figure 1.13 Always wash your hands to prevent ingesting hazardous substances

Welfare facilities

Welfare facilities are things such as toilets, which must be provided by your employer to ensure a safe and healthy workplace. There are several things that your employer must provide to meet welfare standards.

- **Toilets** – the number of toilets provided depends on the number of people who are intended to use them. Males and females can use the same toilets provided there is a lock on the inside of the door. Toilets should be flushable with water or, if this is not possible, with chemicals.

- **Washing facilities** – employers must provide a basin large enough to allow people to wash their hands, face and forearms. Washing facilities must have hot and cold running water as well as soap and a means of drying your hands. Showers may be needed if the work is very dirty or if workers are exposed to **corrosive** and **toxic** substances.

- **Drinking water** – there should be a supply of clean drinking water available, either from a tap connected to the mains or bottled water. Taps connected to the mains need to be clearly labelled as drinking water, and bottled drinking water must be stored in a separate area to prevent **contamination**.

- **Storage or dry room** – every building site must have an area where workers can store the clothes that they do not wear on site such as coats and motorcycle helmets. If this area is to be used as a drying room then adequate heating must also be provided in order to allow clothes to dry.

- **Lunch area** – every site must have facilities that can be used for taking breaks and lunch well away from the work area. These facilities must provide shelter from the wind and rain and must be heated as required. There should be access to tables and chairs, a kettle or urn for boiling water and a means of heating food, for example a microwave.

Key terms

Corrosive – a substance that can damage things it comes into contact with (e.g. material, skin)

Toxic – poisonous

Contamination – when harmful chemicals or substances pollute something (e.g. water)

Safety tips

When placing clothes in a drying room, do not place them directly on to heaters as this can lead to fire

When working in an occupied house, you can make arrangements with the client to use the facilities in their house

Substance abuse

Substance abuse is a general term that mainly covers things such as drinking alcohol and taking drugs.

Taking drugs or inhaling solvents at work is not only illegal, but is also highly dangerous to you and everyone around you. These acts result in reduced concentration problems and can lead to accidents. Drinking alcohol is also dangerous at work; going to the pub for lunch and having just one drink can slow down your reflexes and reduce your concentration.

Did you know?

Substance abuse on a worksite doesn't just endanger yourself, it puts everyone you are working with in danger as well

Although not a form of abuse as such, drugs prescribed by your doctor as well as over-the-counter painkillers can be dangerous. Many of these medicines carry warnings such as 'may cause drowsiness' or 'do not operate heavy machinery'. It is better to be safe than sorry, so always ensure you follow any instructions on prescriptions and, if you feel drowsy or unsteady, stop work immediately.

Health effects of noise

Hearing can be damaged by a range of causes, from ear infections to loud noises. Hearing loss can result from one very loud noise lasting only a few seconds, or from relatively loud noise lasting for hours such as a drill.

The effects of noise damage

To appreciate the damage caused by noise, it helps to first understand how the human ear works.

Did you know?

Noise is measured in decibels (dB). The average person may notice a rise of 3 dB, but with every 3 dB rise, the noise is doubled. What may seem like a small rise is actually very significant

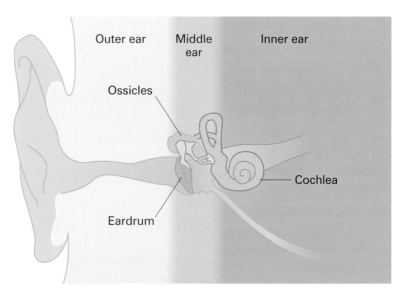

Figure 1.14 Inner workings of the human ear

- When the hairs contained within the ear vibrate they move like grass blowing in the wind. Very loud noises have the same effect on the hairs that a hurricane would have on a field.

- The hairs get blown away and can *never* be replaced.

- The fewer of these hairs you have, the worse your hearing is. This is called 'noise induced hearing loss'.

The damage to hearing can be caused by one of two things:

- **intensity of the noise** – your hearing can be damaged in an instant from an explosive or very loud noise that can burst your ear drum
- **duration of the noise** – noise doesn't have to be deafening to harm you, a quieter noise over a long period, e.g. a 12-hour shift, can also damage your hearing.

Reducing the risks

This can be done in a number of ways:

- **remove** – get rid of whatever is creating the noise
- **move** – locate the noisy equipment away from people
- **enclose** – surround noisy equipment, e.g. with sound-proof material
- **isolate** – move the workers to protected areas.

Even after all of these are considered, PPE may still be required.

Hearing protection

The two most common types of hearing protection aids are ear-plugs and ear defenders. See page 66 for more details.

Hazardous substances

Hazardous substances are major health and safety risks on a construction site. To this end, they need to be handled, stored, transported and disposed of in very specific ways.

Remember

Hearing loss affects young and old

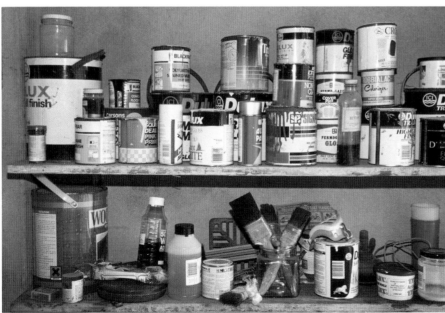

Figure 1.15 Hazardous substances

Safety tip

Not all substances are labelled, and sometimes the label may not match the contents. If you are in any doubt, don't use or touch the substance

To comply with COSHH regulations, eight steps must be followed by the employer:

- **Step 1** – assess the risks to health from hazardous substances used or created by employees' activities.
- **Step 2** – decide what precautions are needed.
- **Step 3** – prevent employees from being exposed to any hazardous substances. If prevention is impossible, the risk must be adequately controlled.
- **Step 4** – ensure control methods are used and maintained properly.
- **Step 5** – monitor the exposure of employees to hazardous substances.
- **Step 6** – carry out health surveillance to ascertain if any health problems are occurring.
- **Step 7** – prepare plans and procedures to deal with accidents such as spillages.
- **Step 8** – ensure all employees are properly informed, trained and supervised.

Identifying a substance that may fall under the COSHH regulations is not always easy, but you can ask the supplier or manufacturer for a COSHH data sheet, outlining the risks involved with a substance. Most substance containers carry a warning sign stating whether the contents are corrosive, harmful, toxic or bad for the environment.

Remember

Always read the manufacturer's label and remember to wear the relevant safety equipment when dealing with hazardous substances

Figure 1.16 Common safety signs for corrosive, toxic and explosive materials

Waste

Many different types of waste material are produced in construction work. It is your responsibility to identify the type of waste you have created and the best way of disposing it.

There are several pieces of legislation that dictate the disposal of waste materials. They include:

- Environmental Protection Act 1990
- Controlled Waste Regulations 1992
- Waste Management Licensing Regulations 1994.

Remember

If you leave material on site when your work is completed you may be discarding them. You are still responsible for this waste material!

Several different types of waste are defined by these regulations:

- household waste – normal household rubbish
- commercial waste – from shops or offices
- industrial waste – from factories and industrial sites.

All waste must be handled properly and disposed of safely. The Controlled Waste Regulations state that only those authorised to do so may dispose of waste and that a record is kept of all waste disposal.

Hazardous waste

Some types of waste such as chemicals or material that is toxic or explosive, are too dangerous for normal disposal and must be disposed of with special care. The Hazardous Waste (England and Wales) Regulations cover this disposal. Examples include:

- asbestos
- used engine oils and filters
- solvents
- pesticides
- lead-based batteries
- oily sludges
- chemical waste
- fluorescent tubes.

If hazardous material is inside a container, the container must be clearly marked and a consignment note completed for its disposal.

Health risks in the workplace

While working in the construction industry, you will be exposed to substances or situations that may be harmful to your health. Some of these health risks may not be noticeable straight away. It may take years for **symptoms** to be noticed and recognised.

Ill health can result from:

- exposure to dust (e.g. asbestos), which can cause eye injuries, breathing problems and cancer
- exposure to solvents or chemicals, which can cause **dermatitis** and other skin problems
- lifting heavy or difficult loads, which can cause back injury and pulled muscles
- bacterial infections such as **leptospirosis**, caused by coming into contact with germs

Key terms

Symptom – a sign of illness or disease (e.g. difficulty breathing, a sore hand or a lump under the skin)

Dermatitis – a skin condition where the affected area is red, itchy and sore

Leptospirosis – an infectious disease that affects humans and animals. The human form is commonly called Weil's disease. The disease can cause fever, muscle pain and jaundice. In severe cases it can affect the liver and kidneys. Leptospirosis is a germ that is spread by the urine of the infected person. It can often be caught from contaminated soil or water that has been urinated on

Key term

Vibration white finger – a condition that can be caused by using vibrating machinery (usually for very long periods of time). The blood supply to the fingers is reduced which causes pain, tingling and sometimes spasms (shaking)

Remember

Activities on site can also damage your body. You could have eye damage, head injury and burns along with other physical wounds

Key term

Kinetic lifting – a way of lifting objects that reduces the risk of injury to the lifter

- exposure to loud noise, which can cause hearing problems and deafness
- exposure to sunlight, which can cause skin cancer
- using vibrating tools, which can cause **vibration white finger** and other problems with the hands
- head injuries, which can lead to blackouts and epilepsy
- cuts, which if infected can lead to disease.

Everyone is responsible for health and safety in the construction industry but accidents and health problems still happen too often. Make sure you do what you can to prevent them.

K5. Safe handling of materials and equipment

Manual handling

Manual handling means lifting and moving a piece of equipment or material from one place to another without using machinery. Lifting and moving loads by hand is one of the most common causes of injury at work. Most injuries caused by manual handling result from years of lifting items that are too heavy, are awkward shapes or sizes, or from using the wrong technique. However, it is also possible to cause a lifetime of back pain with just one single lift.

Poor manual handling can cause injuries such as muscle strain, pulled ligaments and hernias. The most common injury by far is spinal injury. Spinal injuries are very serious because there is very little that doctors can do to correct them and, in extreme cases, workers have been left paralysed.

The Manual Handling Operations Regulations 1992 is the key piece of legislation related to manual handling.

What you can do to avoid injury

The first and most important thing you can do to avoid injury from lifting is to receive proper manual handling training. **Kinetic lifting** is a way of lifting objects that reduces the chance of injury and is covered in more detail in the next section.

Before you lift anything you should ask yourself some simple questions:

- Does the object need to be moved?

- Can I use something to help me lift the object? A mechanical aid such as a forklift or crane or a manual aid such as a wheelbarrow may be more appropriate than a person.

- Can I reduce the weight by breaking down the load? Breaking down a load into smaller and more manageable weights may mean that more journeys are needed, but it will also reduce the risk of injury.

- Do I need help? Asking for help to lift a load is not a sign of weakness and team lifting will greatly reduce the risk of injury.

- How much can I lift safely? The recommended maximum weight a person can lift is 25 kg but this is only an average weight and each person is different. The amount that a person can lift will depend on their physique, age and experience.

- Where is the object going? Make sure that any obstacles in your path are out of the way before you lift. You also need to make sure there is somewhere to put the object when you get there.

- Am I trained to lift? The quickest way to receive a manual handling injury is to use the wrong lifting technique.

Lifting correctly (kinetic lifting)

When lifting any load it is important to keep the correct posture and to use the correct technique.

Get into the correct posture before lifting as follows:

1. feet shoulder width apart with one foot slightly in front of the other
2. knees should be bent
3. back must be straight
4. arms should be as close to the body as possible
5. grip must be firm using the whole hand and not just the fingertips.

The correct technique when lifting is as follows:

1. approach the load squarely facing the direction of travel
2. adopt the correct posture (as above)
3. place hands under the load and pull the load close to your body
4. lift the load using your legs and not your back.

When lowering a load, you must also adopt the correct posture and technique:

- bend at the knees, not the back
- adjust the load to avoid trapping fingers
- release the load.

> **Remember**
>
> Even light loads can cause back problems so when lifting anything, always take care to avoid twisting or stretching

Think before lifting

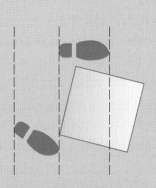

Adopt the correct posture before lifting

Get a good grip on the load

Adopt the correct posture when lifting

Move smoothly with the load

Adopt the correct posture and technique when lowering

Safe handling

Safe manual handling methods are discussed in detail on pages 32–34. When handling any materials or equipment, always think about the health and safety issues involved and remember the manual handling practices explained to you during your induction.

You aren't expected to remember everything but basic common sense will help you to work safely.

- Always wear your safety helmet and boots at work.
- Wear gloves and ear defenders when necessary.
- Keep your work areas free from debris and materials, tools and equipment not being used.
- Wash your hands before eating.
- Use barrier cream before starting work.
- Always use correct lifting techniques.

Ensure you follow instructions given to you at all times when moving any materials or equipment. The main points to remember are:

- always try to avoid manual handling (or use mechanical means to aid the process)
- always assess the situation first to decide on the best method of handling the load
- always reduce any risks as much as possible (e.g. split a very heavy load, move obstacles from your path before lifting)
- tell others around you what you are doing
- if you need help with a load, get it. Don't try to lift something beyond what you can manage.

Did you know?

In 2004/05 there were over 50 000 injuries while handling, lifting or carrying in the UK (Source: HSE)

Working Life

Ahmed and Glynn are unloading bags of plaster from a wheelbarrow. While handling a bag of plaster, Glynn gets a sudden sharp pain in his back and drops the bag. Ahmed tells their supervisor, who comes over to where Glynn is sitting in a great deal of pain.

What do you think should happen next? Glynn will not want to continue working and possibly cause further damage to his back. What should his supervisor do straight away to look after Glynn's well-being?

Could the accident have been prevented? Glynn may not have been working in a safe manner, and there are several important health and safety issues that both Ahmed and Glynn should be made aware of before they carry out another lifting task.

What should their supervisor do in the long term to stop this from happening again? The supervisor will want to make sure that he prevents another accident happening across the whole site. What risk assessments and hazard checks will he need to carry out?

Basic health and safety for tools

All tools are potentially dangerous. You, as an operator of tools, must make sure that all health and safety requirements relating to the tools are always carried out. This will help ensure that you do not cause injury to yourself, and, equally important, to others who may be working around you and to the general public. Make sure you follow any instructions and demonstrations you are given on the use of tools, as well as any manufacturer's instructions provided with purchase of the tools.

Basic rules for handling tools:

- always make sure you use the correct PPE required to use the tool and do the job you are carrying out
- never 'make do' with tools. Using the wrong tool for the job usually breaks health and safety laws
- never play or mess around with a tool regardless of the type, whether it is a hand tool or power tool
- never use a tool you have not been trained to use, especially a power tool.

Power tools

Always treat power tools with respect: they have the potential to cause harm either to the person using them or to others around. All power tools used on site should be regularly tested (PAT tested) by a qualified person. There are several health and safety regulations governing the use of power tools. Make sure that you wear suitable PPE at all times and that power tools are operated safely. In some cases, you must be qualified to use them. Refer to PUWER (Provision and Use of Work Equipment Regulations) 1998 if needed.

On-site transformers are used to reduce the mains voltage from 230 volts to 110 volts. All power tools used should be designed for 110 volts.

As well as the traditional powered tools there are also tools powered by gas or compressed air. Gas-powered tools such as nail guns, also require batteries to operate them. They must be handled carefully similar to other power tools.

Compressed air powered tools such as spray paint systems require an electric powered compressor to operate them. Care must be taken when dealing with these tools. As well as electrical hazards there is the additional danger of working with compressed air. If the air supply is held against the skin it can create air bubbles in your blood stream. This can lead to death.

Remember

Follow the basic health and safety rules regarding use of tools and you (and others) will be safe

Tools are expensive and very important for work so they need to be looked after

Did you know?

PAT stands for 'portable appliance testing'

Special care should be taken with electrical tools.

ALWAYS:

- check plugs and connections (make sure you have the correct fuse rating in the plug)
- inspect all leads to ensure no damage
- check that the power is off when connecting leads
- unwind extension leads completely from the reel to prevent the cable from overheating.

NEVER:

- use a tool in a way not recommended by the manufacturer
- use a tool with loose, damaged or makeshift parts
- lay a driver down while it is still switched on
- use a drill unless the chuck (the part in which the drill bit is held) is tight
- throw the tool onto the ground
- pass the tool down by its lead
- use a drill where it is difficult to see what you are doing or hold the tool tightly
- allow leads to trail in water.

Safety tip

When using power tools, always read the manufacturer's instructions and safety guidelines before use. This will ensure that they are being operated correctly and for the correct purpose

Safe storage and handling of tools and equipment

Hand tools

Hand tools need to be stored safely and securely. Tools such as chisels, saws and craft knives must be stored either in a roll or with a cover over the blade. This is because accidents could happen when people put their hands into tools bags to get something and cut their fingers on a sharp edge.

All tools must be stored in a suitable bag/box that will protect the tools from the elements. With a lot of tools being made from metal components, rust will affect them.

When not in use tools should be securely locked away – theft can occur. It is your responsibility to look after your tools.

Power tools

Power tools should be handled with care. The manufacturer's guidance within the tool's manual will explain the safe handling and storage of the tool. You must follow this guidance.

Generally power tools should be carried by the handle and not the cable. When not in use, the tool should be stored away safely.

Most power tools come in a plastic carry case. They should be kept in this case when not in use and stored in a safe location to protect them from damage and theft. Power tools that have gas-powered cartridges must be stored in an area that is safe and away from sources of ignition to prevent explosion. Used cartridges must be disposed of safely. Pneumatic, hydraulic and air powered tools must also be carried correctly and stored in a way that prevents damage.

Power tools include pressurised painting vessels and equipment and compressed air and hydraulic powered equipment.

Safety tip

NEVER:

- store equipment, cables and plugs in wet areas/outdoors
- store equipment where leads may be damaged (near blades etc.)
- store equipment at height where it may fall on you

ALWAYS:

- store power tools away from children
- allow hot equipment to cool before storing
- unplug and coil lead before storing

Wheelbarrows

Wheelbarrows are generally used where large amounts of material need to be transported over a distance. The best type of wheelbarrow is one specially designed to go through narrow door openings.

Always clean the wheelbarrow out after use. Do not hit the body of the wheelbarrow with a heavy object. Keep tyres inflated, as this will allow for ease of movement. Check that all metal stays are in place.

Do not overload the wheelbarrow as it will put strain on your back and arms. Approach the barrow between the lifting arms and hold the arms at the end, bend your knees and lift. The weight of the material should be over the wheel.

Bricks

Safety tip

Take care and stand well clear of a crane used for offloading bricks on delivery

Most bricks delivered to sites are now pre-packed and banded using either plastic or metal bands to stop the bricks from separating until ready for use. The edges are also protected by plastic strips to avoid damage during moving, usually by forklift or crane. They are then usually covered in shrink-wrapped plastic to protect them from the elements.

On arrival to site bricks should be stored on level ground and stacked no more than two packs high. This is done to prevent over-reaching or collapse, which could result in injury to workers. The bricks should be stored close to where they are required so further movement is kept to a minimum. On large sites they may be stored further away and moved by telescopic lifting vehicles to the position required for use.

If bricks are unloaded by hand they should be stacked on edge in rows, on firm, level and well-drained ground. The ends of the stacks should be bonded and no higher than 1.8 m. To protect the bricks from rain and frost, all stacks should be covered with a tarpaulin or polythene sheets.

Using bricks

Great care should be taken when using the bricks from the packs. Once the banding is cut, the bricks can collapse, which can damage the bricks, especially on uneven ground.

Bricks should be taken from a minimum of three packs and mixed to stop changes in colour. This is because the positioning of the bricks in the kiln can cause slight colour differences:

- the nearer the centre of the kiln, the lighter the colour
- the nearer the edge of the kiln, the darker the colour as the heat is stronger.

If the bricks are not mixed, you could get sections of brickwork in slightly different shades. This is called banding, and this can be clearly made out even by people not working in construction.

Blocks

Blocks are made from concrete, which may be dense or lightweight. Lightweight blocks could be made from a fine aggregate that contains lots of air bubbles. The storage of blocks is the same as for bricks.

Paving slabs

Paving slabs are made from concrete or stone and are available in a variety of sizes, shapes and colours. They are used for pavements and patios, with some slabs given a textured top to improve appearance.

Paving slabs are normally delivered to sites by lorry and crane, some in wooden crates covered with shrink-wrapped plastic, or banded and covered on pallets. They should not be stacked more than two packs high for safety reasons and to prevent damage to the slabs due to weight pressure.

Did you know?

The temperature must be taken into account when laying bricks and blocks as, if it is too cold, this will prevent the cement going off properly

Safety tip

When working with blocks, make sure you always wear appropriate PPE, that is, boots, safety hat, gloves, goggles and facemask (see page 66 for details)

Safety tip

It is good practice to put an intermediate flat stack in long rows to prevent rows from toppling

Figure 1.17 Paving slabs stacked flat

Figure 1.18 Paving slabs on pallet

Figure 1.19 Stacked kerbs

Paving slabs unloaded by hand are stored outside and stacked on edge to prevent the lower ones, if stored flat, from being damaged by the weight of the stack. The stack is started by laying about ten to 12 slabs flat with the others leaning against these. If only a small number of slabs are to be stored, they can be stored flat (since the weight will be less).

Slabs should be stored on firm, level ground with timber bearers below to prevent the edges from getting damaged. This can happen if the slabs are placed on a solid surface.

Kerbs

Kerbs are concrete units laid at the edges of roads and footpaths to give straight lines or curves and retain the finished surfaces. The size of a common kerb is 100 mm wide, 300 mm high and 600 mm long. Path edgings are 50 mm wide, 150 mm high and 600 mm long.

Kerbs should be stacked on timber bearers or with overhanging ends, which provide a space for hands or lifting slings if machine lifting is to be used. When they are stacked on top of each other, the stack must not be more than three kerbs high. To protect the kerbs from rain and frost it is advisable to cover them with a tarpaulin or sheet.

Roofing tiles

Roofing tiles are made from either clay or concrete. They may be machine-made or handmade and are available in a variety of shapes and colours. Many roofing tiles are able to interlock to prevent rain from entering the building. Ridge tiles are usually half round but sometimes they may be angled.

Storage of roofing tiles

Roofing tiles are stacked on edge to protect their 'nibs', and in rows on level, firm, well-drained ground. See Figure 1.21. They should not be stacked any higher than six rows high. The stack should be tapered to prevent them from toppling. The tiles at the end of the rows should be stacked flat to provide support for the rows.

Unit 1001 Safe working practices in construction

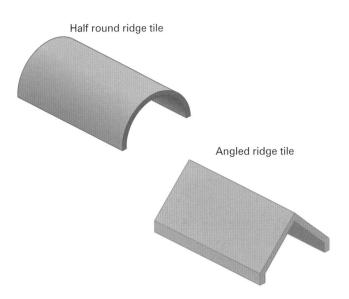

Half round ridge tile

Angled ridge tile

Figure 1.20 Roofing tiles

Ridge tiles may be stacked on top of each other, but not any higher than ten tiles.

To protect roofing tiles from rain and frost before use, they should be covered with a tarpaulin or plastic sheeting.

Storage of rolled materials

Rolled materials, for example damp proof course or roofing felt, should be stored in a shed on a level, dry surface. Narrower rolls may be best stored on shelves but in all cases they should be stacked on end to prevent them from rolling and to reduce the possibility of them being damaged by compression. See Figure 1.22. In the case of bitumen, the layers can melt together under pressure.

Aggregates

Aggregates are granules or particles that are mixed with cement and water to make mortar and concrete. Aggregates should be hard and durable. They should not contain any form of plant life or anything that can be dissolved in water.

Aggregates are classed in two groups:

- fine aggregates are granules that pass through a 5-mm sieve
- coarse aggregates are particles that are retained by a 5-mm sieve.

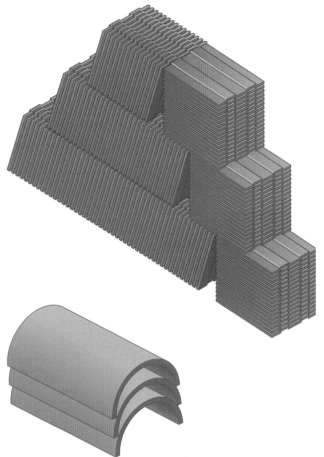

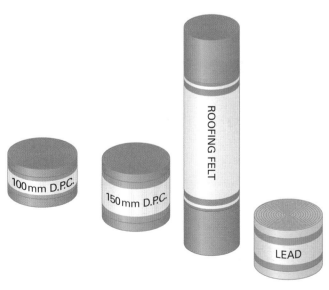

Figure 1.21 Stacks of roofing tiles

Figure 1.22 Rolled materials stored on end

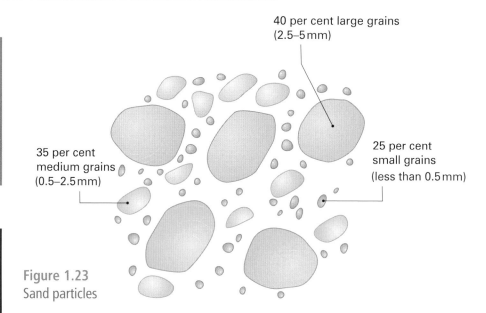

Figure 1.23
Sand particles

40 per cent large grains (2.5–5mm)

35 per cent medium grains (0.5–2.5mm)

25 per cent small grains (less than 0.5mm)

Figure 1.24 Mortar particles

Figure 1.25 Concrete particles

Sand and mortar

The most commonly used fine aggregate is sand. Sand may be dug from pits and riverbeds, or dredged from the sea.

Good mortar should be mixed using 'soft' or 'building' sand. It should be well graded, which means having an equal quantity of fine, medium and large grains.

Concrete

Concrete should be made using 'sharp' sand, which is more angular and has a coarser feel than soft sand, which has more rounded grains.

When concreting, you also need 'coarse aggregate'. The most common coarse aggregate is usually limestone chippings, which are quarried and crushed to graded sizes, 10 mm, 20 mm or even larger.

Storage of aggregates

Aggregates are usually delivered in tipper lorries, although nowadays 1-ton bags are available and may be crane handled. The aggregates should be stored on a concrete base, with a fall to allow for any water to drain away.

To protect aggregates from becoming contaminated with leaves and litter it is a good idea to situate stores away from trees and cover aggregates with a tarpaulin or plastic sheets.

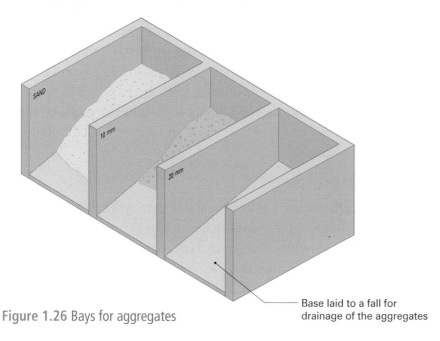

SAND

10 mm

20 mm

Base laid to a fall for drainage of the aggregates

Figure 1.26 Bays for aggregates

Plaster

Plaster is made from gypsum, water and cement or lime. Aggregates can also be added depending on the finish desired. Plaster provides a jointless, smooth, easily decorated surface for internal walls and ceilings.

Gypsum plaster

Gypsum plaster is for internal use and contains different grades of gypsum, depending on the background finish. Browning is usually used as an undercoat on brickwork, but in most cases, a one-coat plaster is used, and on plasterboard, board finish is used.

Cement-sand plaster

This is used for external rendering, internal undercoats and waterproofing finishing coats.

Lime-sand plaster

This is mostly used as an undercoat, but may sometimes be used as a finishing coat.

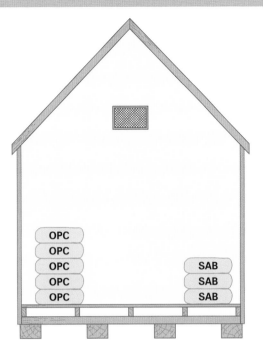

- Dry, ventilated shed

- Stock must be rotated so that old stock is used before new

- Not more than five bags high

- Clear of walls

- Off floor

Figure 1.27 Storage of cement and plaster bags in a shed

Storage of cement and plaster

Both cement and plaster are usually available in 25 kg-bags. The bags are made from multi-wall layers of paper with a polythene liner. Care must be taken not to puncture the bags before use. Each bag, if offloaded manually, should be stored in a ventilated, waterproof shed, on a dry floor on pallets. If offloaded by crane, the bags should be transferred to the shed and the same storage method used.

The bags should be kept clear of the walls, and piled no higher than five bags. It is most important that the bags are used in the same order as they were delivered. This minimises the length of time that the bags are in storage, preventing the contents from setting in the bags, which would require extra materials and cause added cost to the company.

Plasterboard

Plasterboard is a sheet material that comes in various sizes. Some of the larger sheets can be very awkward to carry. Special care must be taken when there is a strong wind as this can catch hold of the sheet and make the handling difficult and dangerous. Ideally two people should be used to transport larger plasterboard sheets.

Plasterboard is a gypsum-based product, so it must be stored in a waterproof area. As it is a sheet material it must also be stored flat and not leant up against a wall. Storing it against a wall will cause the sheet to bow.

Wood and sheet materials

Various types of wood and sheet materials are available. The most common are described below.

Carcassing timber

Carcassing timber is wood used for non-load-bearing jobs such as ceiling and floorboard supports, stud wall partitions and other types of framework. It should normally be stored outside under a covered framework. It should be placed on timber bearers clear of the ground. The ground should be free of vegetation and ideally covered over with concrete. This reduces the risk of absorption of ground moisture, which can damage the timber and cause wet rot.

Remember

The storage racks used to store wood must take account of the weight of the load

Access to the materials being stored is another important consideration

Piling sticks or cross-bearers should be placed between each layer of timber, about 600 mm apart, to provide support and allow air circulation. Tarpaulins or plastic covers can be used to protect the timber from the elements, however care must be taken to allow air to flow freely through the stack. See Figure 1.28.

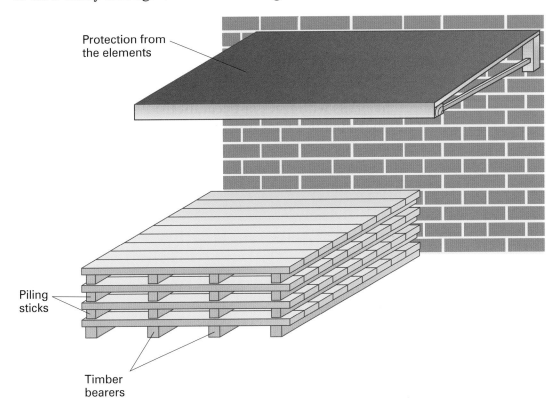

Figure 1.28 Storage of carcassing timber

Joinery grade and hardwoods

These timbers should be stored under full cover wherever possible, preferably in a storage shed. Good ventilation is needed to avoid build-up of moisture through absorption. This type of timber should also be stored on bearers on a well-prepared base.

Plywood and other sheet materials

All sheet materials should be stored in a dry, well-ventilated environment. Specialised covers are readily available to give added protection for most sheet materials. This helps prevent condensation that is caused when non-specialised types of sheeting are used.

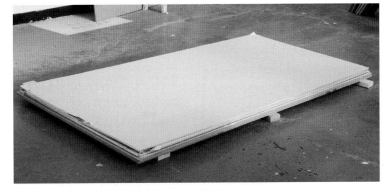

Figure 1.29 Storage of sheet materials

Safety tip

Due to the size, shape and weight of sheet materials, always get help to lift and carry them. If possible, use a purpose-made plywood trolley to transport the load

Find out

What are the PPE requirements when moving sheet materials?

Sheet materials should be stacked flat on timber cross-bearers, spaced close enough together to prevent sagging. Alternatively, where space is limited, sheet materials can be stored on edge in purpose-made racks that allow the sheets to rest against the backboard. There should be sufficient space around the plywood for easy loading and removal. The rack should be designed to allow sheets to be removed from either the front or the ends.

Leaning sheets against walls is not recommended, as this makes them bow. This is difficult to correct.

For sheet materials with faces or decorative sides, the face sides should be placed against each other. This is done to minimise the risk of damage due to friction when they are moved. Different sizes, grades and qualities of sheet materials should be kept separate with offcuts stacked separately from the main stack.

Sheet materials are awkward, heavy and prone to damage so extra care is essential when transporting them. Always ensure that the correct PPE is worn.

Joinery components

Joinery components such as doors or kitchen units, must be stored safely and securely to prevent damage. Doors, windows, frames, etc. should ideally be stored flat on timber bearers under cover, to protect them from exposure to the weather. Where space is limited, they can be stored upright using a rack system (similar to the way sheet materials are stored). However, they must never be leant against a wall. This will bow the door/frame and make it very hard to fit.

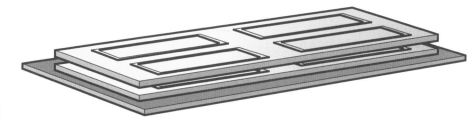

Figure 1.30 Doors should be stacked on a flat surface

Did you know?

Wood and wood-based materials are susceptible to rot if the moisture content is too high, to insect attack and to many other defects such as bowing or warping

Proper seasoning and chemical sprays can prevent defects

Wall and floor units – whether they be kitchen, bedroom or bathroom units – must be stacked on a flat surface, and no more than two units high. Units can be made from porous materials such as chipboard. Therefore, it is vital they are stored inside, preferably in the room where they are to be fitted, to avoid double handling. Protective sheeting should be used to cover units to prevent damage or staining from paints, etc.

Ideally all components and timber products such as architrave should be stored in the room where they are to be fitted. This will allow them to acclimatise to the room and prevent shrinkage or distortion after being fitted. This process is known as 'second seasoning'.

Adhesives

Adhesives are substances used to bond (stick) surfaces together. Because of their chemical nature, there are potentially serious risks connected with adhesives if they are not stored, used and handled correctly.

All adhesives should be stored and used in line with the manufacturer's instructions. This usually involves storing them on shelving, with labels facing outwards, in a safe, secure area (preferably a lockable store room).

The level of risk when using an adhesive depends on the type of adhesive being used. Some of the risks include:

- explosion
- poisoning
- skin irritation
- disease.

As explained on page 29, these types of material are closely controlled by COSHH, which aims to minimise the risks involved with their storage and use.

All adhesives have a recommended **shelf life**. This must be taken into account when storing adhesives to ensure that the oldest stock is stored at the front and used first. Remember to check the manufacturer's guidelines as to how long the adhesive will remain fit for purpose once opened. Poor storage can affect the quality of the adhesives, e.g. loss in adhesive strength and prolongation of the setting time.

Paint and decorating equipment

Oil-based products

Oil-based products such as gloss and varnish should be stored on clearly marked shelves, with their labels turned to the front. They should always be used in date order. New stock should therefore be stored at the back with old stock at the front.

> **Remember**
>
> When storing adhesives, keep the labels facing outwards so that the correct adhesive can be selected

Figure 1.31 Adhesives should be stored according to the manufacturer's instructions

> **Key term**
>
> **Shelf life** – how long something will remain fit for its purpose

> **Did you know?**
>
> Storage instructions can usually be found on the product label or in a manufacturer's information leaflet

Unit 1001 Safe working practices in construction

Figure 1.32 Correct storage of paints

Oil-based products should be **inverted** at regular intervals to stop settlement and separation of the ingredients. They must also be kept in tightly sealed containers to stop the product **skinning**. Storage at a constant temperature will ensure the product retains its desired consistency.

Water-based products

Water-based products, such as emulsions and acrylics, should also be stored on shelves with labels to the front and in date order.

Some water-based products have a very limited shelf life and must be used before their use-by date. As with oil-based products, water-based products keep best if stored at a constant temperature. It is also important to protect them from frost to prevent the water component of the product from freezing.

Powdered materials

Powdered materials a decorator might use include Artex®, fillers, paste and sugar soap.

Large items such as heavy bags should be stored at ground or platform level. Smaller items can be stored on shelves. Sealed containers, e.g. bins, are ideal for loose materials.

Powdered materials can have a limited shelf life and can set in high **humidity** conditions. They must also be protected from frost and exposure to any moisture, including condensation. These types of materials must not be stored in the open air.

Substances hazardous to health

Some substances the decorator will work with are potentially hazardous to health, as they can be **volatile** and highly flammable. COSHH Regulations apply to such materials and describe how they must be stored and handled (see page 29 for general information about COSHH).

Some larger bags of powdered materials are heavier than they appear. Make sure you use the correct manual handling techniques (see page 33).

Decorating materials that might be hazardous to health include spirits (i.e. methylated and white), turpentine (turps), paint thinners and varnish removers. These should be stored out of the way on shelves, preferably in a suitable locker or similar room that meets the requirements of COSHH. The temperature must be kept below 15°C. A warmer environment may cause storage containers to expand and blow up.

The storage of LPG and other highly flammable liquids is covered on pages 21–22.

(see page 29 for general information about COSHH). (see page 33). pages 21–22.

> **Key term**
>
> **Volatile** – a substance that is quick to evaporate (turn into a gas)

K6. Basic working platforms

Working at height

General safety considerations

You should be able to identify potential hazards associated with working at height, as well as hazards associated with equipment. It is essential that access equipment is maintained and checked regularly for any deterioration or faults. These could compromise the safety of the person using the equipment and anyone else in the work area.

Although obviously not as important as people, equipment can also be damaged by the use of faulty access equipment. When maintenance checks are carried out they should be properly recorded. This provides very important information that helps to prevent accidents.

Risk assessment

Before any work is carried out at height, a thorough risk assessment needs to be completed. Your supervisor or someone else more experienced will do this while you are still training. However, it is important that you understand what is involved so that you can carry out risk assessments in future.

Did you know?

Only a fully trained and competent person is allowed to erect any kind of working platform or access equipment. You should therefore not attempt to erect this type of equipment unless this describes you!

Key term

Lanyard – a rope that is used to support a weight

To do a working at height risk assessment properly, a number of questions must be answered:

- How will access and egress to the work area be achieved?
- What type of work is to be carried out?
- How long is the work likely to last?
- How many people will be carrying out the task?
- How often will this work be carried out?
- What is the condition of the existing structure (if any) and the surroundings?
- Is adverse weather likely to affect the work and workers?
- How competent are the workforce and their supervisors?
- Is there a risk to the public and to colleagues?

Duties

Your employer has a duty to provide and maintain safe plant and equipment. This includes scaffold access equipment and systems of work.

You have a duty:

- to comply with safety rules and procedures relating to access equipment
- to understand the hazards in the workplace and report things you consider likely to lead to danger, for example a missing handrail on a working platform
- not to tamper with or modify equipment.

Fall protection

With any task that involves working at height, the main danger to workers is falling. Although scaffolding, etc. should have edge protection to prevent falls, there are certain tasks where edge protection or scaffolding simply can't be used. In these instances some form of fall protection must be in place to:

- prevent the worker falling
- keep the fall distance to a minimum
- ensure the landing point is cushioned.

A variety of fall protection devices are available. The most commonly used ones are:

- harness and **lanyards**
- safety netting
- airbags.

50

Harness and lanyards

Harness and lanyards are a type of **fall-arrest system**. The system works with a harness that is attached to the worker and a lanyard attached to a secure beam/eyebolt. If workers slip, they will fall only as far as the length of the lanyard and will be left hanging, rather than falling to the ground.

Safety netting

Safety netting is also a type of fall-arrest system. It's used mainly on the top floor where there is no higher point to attach a lanyard.

Figure 1.33 A harness and lanyard can prevent a worker from falling to the ground

Figure 1.34 Safety netting is used when working at the highest point

Safety nets are primarily used when decking roofing. They are attached to the joists/beams and are used to catch any worker who may slip or fall. Safety netting is also used on completed buildings where there is a fragile roof.

Airbags

An airbag safety system is a form of soft fall-arrest. It consists of interlinked modular air mattresses. The modules are connected by push connectors and/or flexible couplings and are inflated by a pump-driven fan, which can be electric, petrol, or butane gas powered. As the individual airbags fill with low-pressure air, they expand together to form a continuous protective safety surface, giving a cushioned fall and preventing serious injury.

Figure 1.35 Safety netting can be used under fragile roofs

Did you know?

The British Standards Institute (BSI) was founded in London in 1901. The BSI establishes and maintains standards for practically everything that is used and manufactured in Britain. Items that have been passed by the BSI are stamped with the traditional BSI Kitemark

Safety tip

If any faults are revealed when checking a stepladder, it should be taken out of use, reported to the person in charge, and a warning notice attached to it to stop anyone using it

Figure 1.37 Wooden stepladder

Key term

Stiles – the side pieces of a stepladder into which the steps are set

The system must be kept inflated. If it is run on petrol or gas, it should be checked regularly to ensure that it is still functioning. This system is ideal for short fall jobs, but should not be used where a large fall could occur.

Stepladders and ladders

Stepladders

A stepladder has a prop, which when folded out allows the ladder to be used without having to lean it against something. Stepladders are one of the most frequently used pieces of access equipment in the construction industry and are often used every day. This means that they are not always treated with the respect they demand.

Stepladders are often misused. They should only be used for work that will take a few minutes to complete. When work is likely to take longer, a sturdier alternative needs to be used.

When stepladders are used, the following safety points should be observed.

- Ensure the ground on which the stepladder is to be placed is firm and level. If the ladder rocks or sinks into the ground it should not be used for the work.
- Always open the steps fully.
- Never work off the top tread of the stepladder.
- Always keep your knees below the top tread.
- Never use stepladders to gain additional height on another working platform.
- Always look for the Kitemark (Figure 1.36), which shows that the ladder has been made to BSI standards.

Figure 1.36 BSI Kitemark

A number of other safety points need to be observed depending on the type of stepladder being used.

Wooden stepladder

Before using a wooden stepladder, you should check:

- for loose screws, nuts, bolts and hinges
- that the tie ropes between the two sets of **stiles** are in good condition and not frayed
- for splits or cracks in the stiles
- that the treads are not loose or split.

Never paint any part of a wooden stepladder as this can hide defects, which may cause the ladder to fail during use, causing injury.

Aluminium stepladder

Before using an aluminium stepladder check for damage to stiles and treads to see whether they are twisted, badly dented or loose.

Avoid working close to live electricity supplies as aluminium will conduct electricity.

Fibreglass stepladder

Before using a fibreglass stepladder, check for damage to stiles and treads. Once damaged, fibreglass stepladders cannot be repaired and must be disposed of.

Figure 1.38 Aluminium stepladder

Ladders

A ladder, unlike a stepladder, does not have a prop and has to be leant against something in order for it to be used. Together with stepladders, ladders are one of the most common pieces of equipment used to carry out work at height and to gain access to the work area.

Ladders are also made of timber, aluminium or fibreglass, and require similar checks to stepladders before use.

Pole ladder

These are single ladders and are available in a range of lengths. They are most commonly used for access to scaffolding platforms. Pole ladders are made from timber. They must be stored under cover and flat, supported evenly along their length to prevent them sagging and twisting. They should be checked for damage or defects every time before being used.

Extension ladder

Extension ladders have two or more interlocking lengths. The lengths can be slid together for convenient storage or slid apart to the desired length when in use.

Extension ladders are available in timber, aluminium and fibreglass. Aluminium types are the most favoured as they are lightweight yet strong and available in double and triple extension types. Although also very strong, fibreglass versions are heavy, making them difficult to manoeuvre.

Find out

What are the advantages and disadvantages of different types of stepladder?

Did you know?

Ladders and stepladders should be stored under cover to protect them from damage such as rust or rotting

Figure 1.39 Pole ladder

Figure 1.40 Aluminium extension ladder

Safety tip

Ladders must *never* be repaired once damaged and must be disposed of

Key term

Tie-rods – metal rods underneath the rungs of a ladder that give extra support to the rungs

Safety tip

Always ensure there is a sufficient overlap of rungs when using an extension ladder. A good rule of thumb is to leave a four-rung overlap.

Ladders should only be used as a short-term means of access. If the job is a long one, alternative access equipment should be used.

Did you know?

On average in the UK, every year 14 people die at work falling from ladders, and nearly 1200 suffer major injuries (Source: HSE)

Remember

You must carry out a thorough risk assessment before working from a ladder. Ask yourself, 'Would I be safer using an alternative method?'

Erecting and using a ladder

The following points should be noted when considering the use of a ladder.

- As with stepladders, ladders are not designed for work of long duration. Alternative working platforms (see pages 56–61) should be considered if the work will take longer than a few minutes.
- The work should not require the use of both hands. One hand should be free to hold the ladder.
- You should be able to do the work without stretching.
- You should make sure that the ladder can be adequately secured to prevent it slipping on the surface it is leaning against.

Pre-use checks

Before using a ladder check its general condition. Make sure that:

- no rungs are damaged or missing
- the stiles are not damaged
- no **tie-rods** are missing
- no repairs have been made to the ladder.

In addition, for wooden ladders ensure that:

- they have not been painted, which may hide defects or damage
- there is no decay or rot
- the ladder is not twisted or warped.

Erecting a ladder

Observe the following guidelines when erecting a ladder.

- Ensure you have a solid, level base.
- Do not pack anything under either (or both) of the stiles to level it.
- If the ladder is too heavy to put it in position on your own, get someone to help.
- Ensure that there is at least a four-rung overlap on each extension section.
- Never rest the ladder on plastic guttering as it may break, causing the ladder to slip and the user to fall.
- Where the base of the ladder is in an exposed position, ensure it is adequately guarded so that no one knocks it or walks into it.
- The ladder should be secured at both the top and bottom. The bottom of the ladder can be secured by a second person, however, this person must not leave the base of the ladder while it is in use.

- The angle of the ladder should be a ratio of 1:4 (or 75°). This means that the bottom of the ladder is 1 m away from the wall for every 4 m in height (see Figure 1.41).
- The top of the ladder must extend at least 1 m, or five rungs, above its landing point.

Roof work

When carrying out any work on a roof, a roof ladder or **crawling board** must be used. Roof work also requires the use of edge protection or, where this is not possible, a safety harness.

The roof ladder is rolled up the surface of the roof and over the ridge tiles, just enough to allow the ladder to be turned over and the ladder hook to bear on the tiles on the other side of the roof. This hook prevents the roof ladder sliding down the roof once it is accessed.

Safety tip

The thickness of boards determines the distance between supports

32mm boards can span 1m
38mm boards can span 1.5m
50mm boards can span 2.5m

If boards are spaced over this they will sag or break during use

4 m

1 m

Figure 1.41 Correct angle for a ladder

Key term

Crawling board – a board or platform placed on roof joists that spreads the weight of the worker allowing the work to be carried out safely

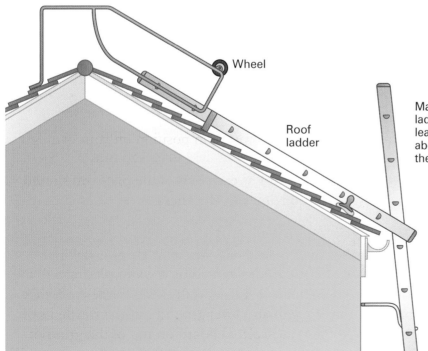

Wheel

Roof ladder

Make sure that the ladder extends at least three rungs above the base of the roof ladder

Safety tip

A scaffold board's maximum span is highlighted on the metal bands on its ends

Use a ladder stand to prevent the access ladder bearing onto the plastic gutter

Figure 1.42 Roof work equipment

Unit 1001

Safe working practices in construction

Trestle platforms

Frames

A-frames

These are most commonly used by carpenters and painters. As the name suggests, the frame is in the shape of a capital A and can be made from timber, aluminium or fibreglass. Two are used together to support a platform (a scaffold or staging board). See Figure 1.43.

Figure 1.43 A-frame trestles with scaffold board

When using A-frames:

- they should always be opened fully and, in the same way as stepladders, must be placed on firm, level ground
- the platform width should be no less than 450 mm
- the overhang of the board at each end of the platform should not be more than four times its thickness.

Steel trestles

These are sturdier than A-frame trestles and are adjustable in height. They are also capable of providing a wider platform than timber trestles – see Figure 1.44. As with the A-frame type, they must be used only on firm and level ground but the trestle itself should be placed on a flat scaffold board on top of the ground. Trestles should not be placed more than 1.2 m apart.

Figure 1.44 Steel trestles with staging board

Platforms

Scaffold boards

To ensure that scaffold boards provide a safe working platform, before using them check that they:

- are not split
- are not twisted or warped
- have no large knots, which cause weakness.

Care should be taken when handling scaffold boards as they can be long and unwieldy. Ideally two people should carry them. It is important to store scaffold boards correctly, that is flat and level, otherwise they will twist or bow. They also need to be covered to prevent damage from rain, which could lead to rot.

Staging boards

These are designed to span a greater distance than scaffold boards and can offer a 600 mm wide working platform. They are ideal for use with trestles.

Hop-ups

Also known as step-ups, hop-ups are ideal for reaching low-level work that can be carried out in a relatively short period of time. A hop-up needs to be of sturdy construction and have a base of not less than 600 mm by 500 mm. Hop-ups have the disadvantage that they are heavy and awkward to move around.

Safety tip

Do not use items as hop-ups that are not designed for this purpose (e.g. milk crates, stools or chairs). They are usually not very sturdy and can't take the weight of someone standing on them. This may result in falls and injury

Scaffolding

Tubular scaffold is the most commonly used type of scaffolding within the construction industry. There are two types of tubular scaffold:

- Independent scaffold – free-standing scaffold that does not rely on any part of the building to support it (although it must be tied to the building to provide additional stability).
- Putlog scaffold – scaffolding that is attached to the building via the entry of some of the poles into holes left in the brickwork by the bricklayer. The poles stay in position until the construction is complete and give the scaffold extra support.

No one other than a qualified **carded scaffolder** is allowed to erect or alter scaffolding. Although you are not allowed to erect or alter this type of scaffold, you must be sure it is safe before you work on it. Ask yourself the following questions to assess the condition and suitability of the scaffold before you use it:

- Are there any signs attached to the scaffold which state that it is incomplete or unsafe?
- Is the scaffold overloaded with materials such as bricks?
- Are the platforms cluttered with waste materials?
- Are there adequate guard rails and scaffold boards in place?
- Does the scaffold actually 'look' safe?
- Is there correct access to and from the scaffold?
- Are the various scaffold components in the correct place (see Figure 1.45)?

Key term

Carded scaffolder – someone who holds a recognised certificate showing competence in scaffold erection

Did you know?

It took 14 years of experimentation to finally settle on 48 mm as the diameter of most tubular scaffolding poles

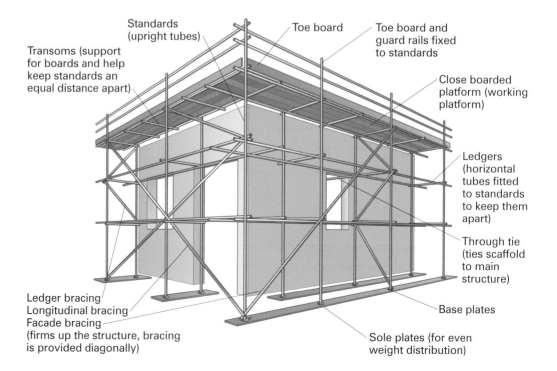

Figure 1.45
Components of a tubular scaffolding structure

Standards (upright tubes)

Transoms (support for boards and help keep standards an equal distance apart)

Toe board

Toe board and guard rails fixed to standards

Close boarded platform (working platform)

Ledgers (horizontal tubes fitted to standards to keep them apart)

Through tie (ties scaffold to main structure)

Base plates

Sole plates (for even weight distribution)

Ledger bracing
Longitudinal bracing
Facade bracing (firms up the structure, bracing is provided diagonally)

● Have the correct types of fittings been used (see Figure 1.46)?

Right angle coupler – load bearing; used to join tubes at right angles

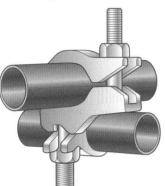

Universal coupler – load bearing; also used to join tubes at right angles

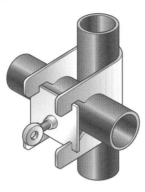

Swivel coupler – load bearing; used to join tubes at various angles, e.g. diagonal braces

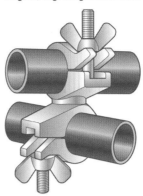

Adjustable base plate or base plate used at the base of standards to allow even weight distribution

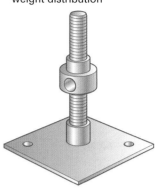

Figure 1.46 Types of scaffold fittings

If you have any doubts about the safety of scaffolding, report them. You could very well prevent serious injury or even someone's death.

Working Life

Sanjit has been asked by a client to take a look at all the fascia boards on a two-storey building. Depending on the condition of the fascia boards, they will need either repairing or replacing. The job will probably take Sanjit between two and six hours, depending on what he has to do.

- What types of scaffolding do you think might be suitable for Sanjit's job?
- Can you think of anything not listed below that Sanjit will need to consider while he prepares for and carries out this task?
  He will need to think about things such as access and egress points, whether the area will be closed off to the public, how long he will need to work at height, etc.
- Take a look through the information on the following pages on types of scaffold. What type of scaffold do you think Sanjit should use for this task?

Mobile tower scaffolds

Mobile tower scaffolds are so called because they can be moved around without being dismantled. Lockable wheels make this possible and they are used extensively throughout the construction industry by many different trades. A tower can be made from either traditional steel tubes and fittings or aluminium, which is lightweight and easy to move. The aluminium type of tower is normally specially designed and is referred to as a 'proprietary tower'.

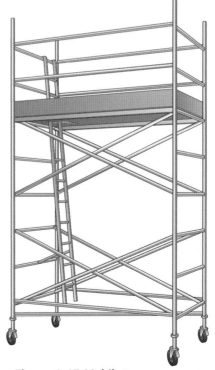

Figure 1.47 Mobile tower scaffold

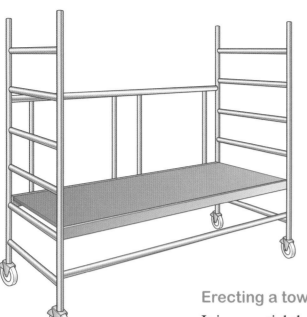

Figure 1.48 Low tower scaffold

Low towers

These are a smaller version of the standard mobile tower scaffold and are designed specifically for use by one person. They have a recommended working height of no more than 2.5 m and a safe working load of 150 kg. They are lightweight and easily transported and stored.

These towers require no assembly other than the locking into place of the platform and handrails. However, you still require training before you use one and you must ensure that the manufacturer's instructions are followed when setting up and working from this type of platform.

Erecting a tower scaffold

It is essential that tower scaffolds are situated on a firm and level base. The stability of any tower depends on the height in relation to the size of the base:

- For use inside a building, the height should be no more than three and a half times the smallest base length.
- For outside use, the height should be no more than three times the smallest base length.

The height of a tower can be increased, provided the area of the base is increased **proportionately**. The base area can be increased by fitting outriggers to each corner of the tower.

For mobile towers, the wheels must be in the locked position while they are in use and unlocked only when they are being repositioned.

There are several important points you should observe when working from a scaffold tower.

- Any working platform above 2 m high must be fitted with guard rails and toe boards. Guard rails may also be required at heights of less than 2 m if there is a risk of falling onto potential hazards below, for example reinforcing rods. Guard rails must be fitted at a minimum height of 950 mm.
- If guard rails and toe boards are needed, they must be positioned on all four sides of the platform.
- Any tower higher than 9 m must be secured to the structure.
- Towers must not exceed 12 m in height unless they have been specifically designed for that purpose.
- The working platform of any tower must be fully boarded and be at least 600 mm wide.

- If the working platform is to be used for materials, the minimum width must be 800 mm.
- All towers must have their own access and this should be by an internal ladder.

The dangers of working at height

While working at height there are a number of dangers that need to be identified. The obvious danger is falling from the height, which can result in serious injury, but there are also additional dangers in working at height that are not present when working at ground level.

Although good housekeeping (see page 20) is important while working at ground level to prevent slips and trips, it is *vital* when working at height. Not only are you at added risk, but materials and tools that are left on a working platform can be knocked off the platform onto people working below. There is a risk of causing serious head injuries to those below – and not just the workforce, as in some instances the working platform may be in an area that involves the general public.

When working in a public area, the public must be protected from hazards by way of barriers around the work area. You must also ensure that the sides of the working platform are sealed off to prevent any materials or other objects from falling.

K7. Working with electricity

Electricity is a killer. One of the main problems with electricity is that it is invisible. You don't even have to be working with an electric tool to be electrocuted. You can get an electric shock:

- working too close to live overhead cables
- plastering a wall with electric sockets
- carrying out maintenance work on a floor
- drilling into a wall.

However, not all electric shocks are fatal – they can also cause injuries such as burns and problems with your muscles and heart.

A common error is to think that the level of voltage is directly related to the level of injury or danger of death. However, a small shock from static electricity may contain thousands of volts but has very little current behind it.

Around 30 workers a year die from electricity-related accidents, with over 1000 more being seriously injured (Source: HSE)

Voltages

There are two main types of voltage in use in the UK: 230 V and 110 V. The standard UK power supply is 230 V and this is what all the sockets in your house are. On construction sites, 230 V has been deemed as unsafe and 110 V must be used here. The 110 V is identified by a yellow cable and different style plug. It works from a transformer which converts the 230 V to 110 V.

When working within domestic dwellings where 230 V is the standard power source, a portable transformer should be used. If this is not possible then residual current devices (RCD) should be used.

Contained within the wiring there should be three wires: the live and neutral, which carry the alternating current, and the earth wire, which acts as a safety device. The three wires are colour-coded so that all electrical installations are standardised and any person needing to do work can easily identify which wire is which. Current coding complies with European colours as follows:

- live – brown
- neutral – blue
- earth – yellow and green.

Some older properties will have the following (older) colour-coding:

- live – red
- neutral – black
- earth – yellow and green.

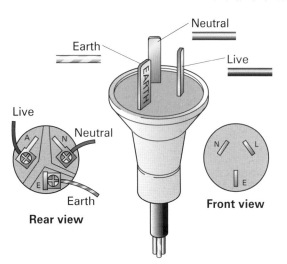

Figure 1.49 Colour coding of the wires in a 230 V plug

Figure 1.50 A 110 V plug

Precautions to take to prevent electric shocks

NEVER:

- carry electrical equipment by the cable
- remove plugs by pulling on the lead
- allow tools to get wet. If they do, get them checked before use.

ALWAYS:

- check equipment, leads and plugs before use. If you find a fault don't use the equipment and tell your supervisor immediately
- keep the cable off the ground where possible to avoid damage/trips
- avoid damage to the cable by keeping it away from sharp edges
- keep the equipment locked away and labelled to prevent it being used by accident

- use cordless tools where possible
- follow instructions on extension leads.

Dealing with electric shocks

In helping the victim of an electric shock, the first thing you must do is disconnect the power supply – if it is safe to do and will not take long to find. Touching the power source may put you in danger.

- If the victim is in contact with something portable such as a drill, attempt to move it away using a non-conductive object such as a wooden broom.

- Time is precious and separating the victim from the source can prove an effective way to speed the process.

- Don't attempt to touch the affected person until they are free and clear of the supplied power. Be especially careful in wet areas such as bathrooms – most water will conduct electricity and electrocuting yourself is also possible.

People 'hung up' in a live current flow may think they are calling out for help but most likely no sound will be heard from them. When the muscles contract under household current (most electrocutions happen from house current at home), the person affected will appear in 'locked-up' state, unable to move or react to you.

- Using a wooden object, swiftly and strongly knock the person free, trying not to injure them, and land them clear of the source.

- The source may also be lifted or removed, if possible, with the same wooden item. This is not recommended on voltages that exceed 500 V.

First aid procedures for an electric shock victim

- Check to see if you are alone. If there are other people around, instruct them to call an ambulance immediately.

- Check for a response and breathing.

- If the area is safe for you to be in, and you have removed the object or have cut off its power supply, talk to the person to see if they are conscious. At this stage, do not touch the victim.

- Check once again to see if the area is safe. If you are satisfied that it is safe, start resuscitating the victim. If you have no first aid knowledge, call 999 for an ambulance.

Safety tip

Don't attempt going near a victim of an electric shock without wearing rubber or some form of insulated sole shoes; bare or socked feet will allow the current to flow to ground through your body as well

Tyrone and Macy are knocking down a small wall in an old block of flats. The wall has an electric socket in it. Tyrone says that they should switch off the power at the mains, disconnect the socket and put some tape over the wires. Macy isn't sure, but they find the mains switch in the flat and switch off the power. Tyrone removes the socket cover and suddenly there is a bang. Tyrone is thrown backwards.

- What do you think has happened?
- How could this have happened?
- What should have been done?
- What should Macy do now?

K8. Using appropriate PPE

Personal protective equipment is the name for clothes and other wearable items that form a line of defence against accidents or injury. PPE is not the only way of preventing accidents or injury. It should be used with all the other methods of staying healthy and safe in the workplace (equipment, training, regulations and laws, etc.).

Maintaining and storing PPE

It is important that PPE is well maintained. The effectiveness of the protection it offers will be affected if the PPE is damaged in any way. Maintenance may include:

- cleaning
- examination
- replacement
- repair and testing.

The wearer may be able to carry out simple maintenance (such as cleaning), but more intricate repairs must only be carried out by a competent person. The costs associated with the maintenance of PPE are the responsibility of the employer.

Where PPE is provided, adequate storage facilities for PPE must also be provided for when it is not in use, unless the employee may take PPE away from the workplace (e.g. footwear or clothing).

Accommodation may be simple (e.g. pegs for waterproof clothing or safety helmets) and it needs not be fixed (e.g. a case for safety glasses or a container in a vehicle). Storage should be adequate to protect the PPE from contamination, loss, damage, damp or

Remember

PPE only works properly if it is being used and used correctly!

The main pieces of legislation that govern the use of PPE are:

- Control of Substances Hazardous to Health 2002
- Provision and Use of Work Equipment Regulations (1992 and 1998)
- Personal Protective Equipment at Work Regulations 1992

sunlight. Where PPE may become contaminated during use, storage should be separate from any storage provided for ordinary clothing.

PPE must be maintained regularly.

PPE should be 'CE' marked. This will indicate that is complies with the requirements of the Personal Protective Equipment Regulations 2002. The CE marking shows that the PPE meets safety requirements. In some cases it may have been tested and certified by an independent body.

PPE must be supplied by your employer free of charge. You have the responsibility as an employee to look after it and use it whenever it is required. See page 10 for the conditions laid out for the protection and maintenance of PPE by the regulations. The HSE website (www.hse.gov.uk) also gives information on maintenance and use of PPE.

The possible consequences of not using PPE can be serious and cause long-term health problems. The health problems and their consequences are described on page 31.

Types of PPE

Different jobs require different types of PPE – the protection needed while using a circular saw is different from the protection needed building a gable end. Some body parts need more protection than others! Each piece of PPE must be suitable for the job and used properly.

Head protection

There are several different types of head protection; the one most commonly used in construction is the safety helmet (or hard hat). This is used to protect the head from falling objects and knocks and has an adjustable strap to ensure a snug fit. Some safety helmets come with attachments for ear defenders or eye protection. Safety helmets are meant to be worn directly on the head and must not be worn over any other type of hat.

Figure 1.51 A safety helmet

Eye protection

Eye protection is used to protect the eyes from dust and flying debris. The three main types are:

- safety goggles – made of a durable plastic and used when there is a danger of dust getting into the eyes or a chance of impact injury

Figure 1.52 Safety goggles

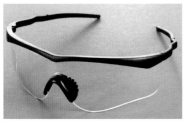

Figure 1.53 Safety spectacles

Figure 1.55 Earplugs

Figure 1.56 Ear defenders

- safety spectacles – these are also made from a durable plastic but give less protection than goggles. This is because they don't fully enclose the eyes and only protect from flying debris

- facemasks – again made of durable plastic, facemasks protect the entire face from flying debris. They do not, however, protect the eyes from dust.

Foot protection

Safety boots or shoes are used to protect the feet from falling objects and to prevent sharp objects such as nails from injuring the foot. Safety boots should have a steel toe-cap and steel mid-sole.

Figure 1.54 Safety boots

Hearing protection

Hearing protection is used to prevent damage to the ears caused by very loud noise. There are several types of hearing protection available, but the two most common types are earplugs and ear defenders.

- **Earplugs** – these are small fibre plugs that are inserted into the ear and used when the noise is not too severe. Before inserting earplugs, make sure that your hands are clean. Never use plugs that have been used by somebody else.

- **Ear defenders** – these are worn to cover the entire ear and are connected to a band that fits over the top of the head. They are used when there is excessive noise and must be cleaned regularly.

Respiratory protection

Respiratory protection is used to prevent the worker from breathing in any dust or fumes that may be hazardous. The main type of respiratory protection is the dust mask.

Dust masks are used when working in a dusty environment and are lightweight, comfortable and easy to fit. They should be worn by only one person and must be disposed of at the end of the working day.

Figure 1.57 A respiratory system

Hand protection

There are several types of hand protection and the correct type must be used for the task at hand. To make sure you are wearing the most suitable type of glove for the task, you need to look first at what is going to be done and then match the type of glove to that task.

Figure 1.58 Safety gloves

Skin and sun protection

Another precaution you can take is ensuring that you wear barrier cream. This is a cream used to protect the skin from damage and infection. Don't forget to ensure that your skin is protected from the sun with a good sunscreen, and make sure your back, arms and legs are covered by suitable clothing.

Whole body protection

The rest of the body also needs protecting when working on site. This will usually involve wearing either overalls or a high-visibility jacket.

High-visibility jackets are essential whenever you are on site or working near traffic. They ensure that the person wearing them is clearly visible at all times. This helps to avoid accidents by making the wearer easier to avoid.

Overalls provide protection from dirt and the possibility of minor cuts. In wet conditions you may also need to use waterproof or thermal clothing. Some circumstances will require chemical-resistant clothing.

Knee pads can be worn by workers who spend a lot of time kneeling, such as carpet fitters. Paper overalls or paper boiler suits can be worn for such tasks as insulating a loft where irritant fibres may be a problem.

Figure 1.59 High-visibility jacket

Figure 1.60 Overalls

Figure 1.61 The triangle of fire

Unit 1001 Safe working practices in construction

K9. Fire and emergency procedures

Fires can start almost anywhere and at any time, but a fire needs three things to burn:

- fuel
- heat
- oxygen.

Together these elements are known as 'the triangle of fire'. If any one of the three ingredients is missing, fire cannot burn. Remove one side of the triangle, and the fire will be extinguished.

If it can consume all three ingredients from the triangle, a fire will spread. Cutting off the fire's access to fuel, heat, or oxygen will stop the spread of the fire.

Fire moves from area to area either by burning the fuel along the way – paper or wood shavings, for example – or through the direct transfer of heat. If a burning piece of plywood is leaning against another piece of plywood, both pieces of plywood will eventually go up in flames, causing the fire to spread.

Fires can be classified according to the type of material that is involved:

- Class A – wood, paper, textiles, etc.
- Class B – flammable liquids, petrol, oil, etc.
- Class C – flammable gases, LPG, propane, etc.
- Class D – metal, metal powder, etc.
- Class E – electrical equipment.

Fire-fighting equipment

There are several types of fire-fighting equipment such as fire blankets and fire extinguishers. Each type is designed to be the most effective at putting out a particular class of fire. Some should never be used in certain types of fire.

Remember

- Remove the fuel – without anything to burn, the fire will go out
- Remove the heat and the fire will go out
- Remove the oxygen and the fire will go out – without oxygen, a fire won't even ignite

Figure 1.62 Water fire extinguisher

Figure 1.63 Foam fire extinguisher

Figure 1.64 Carbon dioxide (CO_2) extinguisher

Figure 1.65 Dry powder extinguisher

Fire extinguishers

A fire extinguisher is a metal canister containing a substance that can put out a fire. There are several different types and it is important that you learn which type should be used on specific classes of fires. This is because if you use the wrong type, you may make the fire worse or risk severely injuring yourself.

Fire extinguishers are now all one colour (red) but they have a band of colour which shows what substance is inside.

Water

The coloured band is red and this type of extinguisher can be used on Class A fires. Water extinguishers can also be used on Class C fires to cool the area down.

Foam

The coloured band is cream and this type of extinguisher can be used on Class A fires. A foam extinguisher can also be used on a Class B fire if the liquid is not flowing, and on a Class C fire if the gas is in liquid form.

Carbon dioxide (CO_2)

The coloured band is black and the extinguisher can be used primarily on electrical fires. However, as well as Class E, it can also be used on Class A, B and C fires.

Find out

What fire risks are there in the construction industry? Think about some of the materials (fuel) and heat sources that could make up two sides of 'the triangle of fire'

Safety tip

A water fire extinguisher should *never* be used to put out an electrical or burning fat/oil fire. This is because electrical current can carry along the jet of water back to the person holding the extinguisher, electrocuting them. Putting water on to burning fat or oil will make the fire worse as the fire will 'explode', potentially causing serious injury

Figure 1.66 A fire blanket

Dry powder

The coloured band is blue and this type of extinguisher can be used on all classes of fire, but most commonly on electrical and liquid fires. The powder puts out the fire by smothering the flames.

Fire blankets

Fire blankets are normally found in kitchens or canteens as they are good at putting out cooking fires. They are made of a fireproof material and work by smothering the fire and stopping any more oxygen from getting to it, thus putting it out. A fire blanket can also be used if a person is on fire.

It is important to remember that when you put out a fire with a fire blanket, you must take extra care as you will have to get quite close to the fire.

What to do in the event of a fire

During your induction to any workplace, you will be made aware of the fire procedure as well as where the fire assembly points (also known as 'muster points') are and what the alarm sounds like.

All muster points should be clearly indicated by signs, and a map of their location clearly displayed in the building. On hearing the alarm you must stop what you are doing and make your way to the nearest muster point. This is so that everyone can be accounted for. If you do not go to the muster point or if you leave before someone has taken your name, someone may risk their life to go back into the fire to get you.

When you hear the alarm, you should not stop to gather any belongings and you must not run. If you discover a fire, you must only try to fight the fire if it is blocking your exit or if it is small. Only re-enter the site or building when you have been given the all-clear.

Every building and organisation will have its own unique fire evacuation procedures and practices. Make sure that you are familiar with the procedures in your workplace so that you will know what to do in the event of an evacuation. Fire drills should be part of every organisation's routine, to ensure that the procedures and practices required in the case of a fire are well known to everyone in the building.

You will see safety signs in many parts of the workplace.

Safety tips

Fire and smoke can kill in seconds, so think and act clearly, quickly and sensibly

Evacuation procedures are not just to protect your safety, but everyone's safety. If you do not follow the correct procedures and go to the correct assembly areas, other people may risk their own lives trying to find you

K10. Safety signs and notices

Uses of safety signs

Safety signs are used to:

- warn of any hazards
- prevent accidents
- explain where things are
- tell you what to do – or not do! – in certain areas.

Figure 1.67 A prohibition sign

Types of safety sign

There are several different types of safety sign, and they have different purposes.

- **Prohibition signs** – these tell you that something *must not* be done. They always have a white background and a red circle with a red line through it.
- **Mandatory signs** – these tell you that something *must* be done. They are also circular, but have a white symbol on a blue background.
- **Warning signs** – these signs are there to alert you to a specific hazard. They are triangular and have a yellow background and a black border.

Figure 1.68 A mandatory sign

- **Safe condition signs (often called information signs)** – these give you useful information like the location of things (e.g. a first aid point). They can be square or rectangular and are green with a white symbol.

Most signs consist of only the symbols that let you know what they are saying. Others have some words as well. For example, a no-smoking sign might have a cigarette in a red circle with a red line crossing through the cigarette and the words 'No smoking' underneath.

Remember

Make sure you take notice of safety signs in the workplace – they have been put up for a reason!

Figure 1.69 A warning sign

Figure 1.70 An information sign

Figure 1.71 A safety sign with symbol and words

FAQ

How do I find out which safety legislation is relevant to my job?

Ask your employer or manager, or contact the HSE at www.hse.gov.uk.

When do I need to do a risk assessment?

A risk assessment should be carried out if there is any chance of an accident happening as a direct result of the work being done. To be on the safe side, you should make a risk assessment before starting each task.

Do I need to read and understand every regulation?

No, it is part of your employer's duty to ensure that you are aware of what you need to know.

Do I have to attend every tool box talk?

No, you only need to attend the tool box talks relevant to you. If you are unsure, or think you have missed a tool box talk, discuss it with your supervisor.

What do I need to do if my PPE is damaged?

You need to inform your employer immediately so that you can have the PPE replaced. Damaged PPE will not offer sufficient protection.

Check it out

1. Name five pieces of health and safety legislation that affect the construction industry and give a brief explanation of what they do.
2. What is the HSE? Give a brief explanation of its role.
3. Sketch the triangle of fire and explain how the parts of the triangle relate to each other.
4. Describe, with sketches, the major warning signs and explain what they do.
5. Name the six different types of PPE and give an example of each.
6. Describe the key items that should be included in a first aid kit.
7. Describe the purpose of a tool box talk.
8. Describe three key things you should never do with a portable power tool.
9. Using a material that is familiar to you, explain, with the aid of sketches if necessary, how that material should be stored.
10. Describe how a wooden stepladder should be checked before use.
11. When should a trestle platform be used? What two types of board can be used as a platform with a trestle frame?

Getting ready for assessment

Unit 1001 Safe working practices in construction

The information contained in this chapter, as well as continued health and safety good practice throughout your training, will help you with preparing for both your end of unit test and the diploma multiple-choice test. It will also help you to understand the dangers of working in the construction industry. Wherever you work in the construction industry, you will need to understand the dangers involved. You will also need to know the safe working practices for the work required for your synoptic practical assignments.

Your college or training centre should provide you with the opportunity to practise these skills, as part of preparing for the test.

You will need to know about and understand the dangers that could arise, and precautions that can be taken, for:

- the safety rules and regulations
- knowing accident and emergency procedures
- identifying hazards on site
- health and hygiene
- safe handling of materials and equipment
- working at heights
- working with electricity
- using personal protective equipment (PPE)
- fire and emergency procedures
- safety signs.

You will need to apply the things you have learnt in this chapter to the actual work you will be carrying out in the synoptic test, and in your professional life. For example, with learning outcome 6 you have seen why basic working platforms are used and the good practice you should use when working on these platforms. You have also seen the different parts of ladders and scaffolding and identified the dangers of working at height. You will now need to use this knowledge yourself when you are working, by using access equipment to the correct legislation and safeguarding your health, through using the correct PPE. You will also need to use your understanding of how PPE should be stored to maintain it in perfect condition.

Before you start work you should always think of a plan of action. You will need to know the clear sequence of operations for the practical work that is to be constructed to be sure you are not making mistakes as you work and that you are working safely at all times.

Your speed in carrying out these tasks in a practice setting will also help to prepare you for the time set for the test. However you must never rush the test! This is particularly important with health and safety, as you must always make sure you are working safely. Make sure throughout the test that you are wearing appropriate and correct PPE and using tools correctly.

This chapter has explained the dangers you may face when working. Understanding these dangers and the precautions that can be taken to help prevent them, will not only aid you in your training but will help you remain safe and healthy throughout your working life.

Good luck!

CHECK YOUR KNOWLEDGE

1 Legislation is:
- **a** a law that must be complied with.
- **b** a guide to tell you what to do.
- **c** a code of practice.
- **d** not your responsibility.

2 Accidents are caused by:
- **a** following all instructions carefully.
- **b** taking care of yourself and others.
- **c** hurrying and not paying attention.
- **d** nothing – accidents just happen.

3 Manual-handling injuries can be caused by:
- **a** lifting items that are too heavy.
- **b** lifting an item once.
- **c** lifting an item repetitively.
- **d** all of the above.

4 Which regulation deals with lifting components?
- **a** The Work at Height Regulations.
- **b** The Manual Handling Operations Regulations.
- **c** The PPE Regulations.
- **d** The Electricity at Work Regulations.

5 With regards to PPE, the employer must:
- **a** supply you with it.
- **b** not charge you for it.
- **c** ensure you wear it.
- **d** do all of the above.

6 A sign that is circular with a white background and a red circle around the edge with a red line through it is:
- **a** a warning sign.
- **b** a mandatory sign.
- **c** a prohibition sign.
- **d** an information sign.

7 Which of the following regulations deals with chemicals?
- **a** Control of Substances Hazardous to Health.
- **b** Health and Safety at Work Act.
- **c** Provision and Use of Work Equipment Regulations.
- **d** Control of Noise at Work Regulations.

8 Who should carry out risk assessments?
- **a** no one.
- **b** everyone.
- **c** an untrained supervisor.
- **d** a trained supervisor.

9 Welfare facilities must include a:
- **a** bathing area.
- **b** sleeping area.
- **c** lunch area.
- **d** TV room.

10 What voltage is the plug shown below?
- **a** 230 V
- **b** 110 V
- **c** 240 V
- **d** 120 V

Unit 1002

Information, quantities and communicating with others

The construction industry contains a great deal of information that you must fully understand to meet the needs of a project. This information can be presented and stored in many ways. Some of the most important information relates to the start of a project, and tells you the type of construction you will be working on, what it will be used for and what you will need to build it.

When working with information, you will also be working with other people. You will need to make sure that you use the correct method to communicate different types of information to other people.

This unit contains material that supports the five generic units. It will also support your completion of scaled drawings throughout all TAP units.

This unit will cover the following learning outcomes:

- Interpreting building information
- Determining quantities of materials
- Communicating information in the workplace

When reading and understanding the text in this unit, you are practising several functional skills.

FE 1.2.1 - Identifying how the main points and ideas are organised in different texts.

FE 1.2.2 – Understanding different texts in detail.

FE 1.2.3 – Read different texts and take appropriate action, e.g. respond to advice/instructions.

If there are any words or phrases you do not understand, use a dictionary, look them up using the internet or discuss with your tutor.

Did you know?

Before doing work on certain public sector jobs e.g. working for the Ministry of Defence or certain government buildings, employees may have to sign and work to certain confidentiality clauses. This means that the information they work with must be kept secret

K1. Interpreting building information

Building information is often presented as written documents. These can be:

- paperwork
- forms
- plans
- diagrams.

To understand all of these documents you need to know what they are used for in a building project.

Diagrams and plans are not drawn to life-size, but to scales. To be able to use diagrams, you need to understand how scales work and what they tell you about the final building.

In this section you will learn about:

- document security
- how to use scales and symbols
- specifications, drawings and schedules.

Document security

A lot of the information we will be looking at in this unit is vitally important for smooth and safe working on a site. Therefore it is important to store documents safely, securely and correctly. You need to understand this before you begin to work with documents.

Some documents may contain sensitive information such as addresses, National Insurance numbers, etc. The Data Protection Act (1998) states that any company that holds personal information must make sure that it is secure and is only used for the purpose that it was provided for. This means that your employer, who will have information such as your bank details, must ensure all of this is kept secure.

Other documents such as drawings should also be kept secure, especially during the duration of the job. This is because these are a permanent record, both of what is happening and what has yet to happen. It is also considered good practice to keep records even after a job has been completed. This is because these can be used for reference at a later date and can be invaluable when estimating the cost of other similar jobs.

One problem with record-keeping is obviously storage, and large companies have thousands of documents. However, storage is becoming a lot easier as we can now create and save documents on computers. We can even scan hard copies and save the original documents electronically too.

Scales, symbols and abbreviations

All building plans are drawn to scales by using symbols and abbreviations. To draw a building on a drawing sheet, its size must be reduced. This is called a scale drawing.

Using scales

The scales that are preferred for use in building drawings are shown in Table 2.1.

Type of drawing	Scales
Block plans	1:2500, 1:1250
Site plans	1:500, 1:200
General location drawings	1:200, 1:100, 1:50
Range drawings	1:100, 1:50, 1:20
Detail drawings	1:10, 1:5, 1:1
Assembly drawings	1:20, 1:10, 1:5

Table 2.1 Preferred scales for building drawings

These scales mean that, for example, on a block plan drawn to 1:2500, 1 mm on the plan would represent 2500 mm (or 2.5 m) on the actual building. Some other examples are:

- on a scale of 1:50, 10 mm represents 500 mm
- on a scale of 1:100, 10 mm represents 1000 mm (1.0 m)
- on a scale of 1:200, 30 mm represents 6000 mm (6.0 m).

Accuracy of drawings

Printing or copying of drawings introduces variations that affect the accuracy of drawings. Hence, although measurements can be read from drawings using a rule with common scales marked (Figure 2.1), you should work to written instructions and measurements wherever possible.

Remember

A scale is merely a convenient way of reducing a drawing in size

Functional skills

In addition to your reading skills, you will be using and practising several functional skills in this unit.

FM 1.1.1 – Identifying and selecting mathematical procedures.

FM 1.2.1a – Using appropriate mathematical procedures

FM 1.2.2b – Interpreting information from various sources – such as the diagrams, tables and charts throughout this unit.

Find out

With a little practice, you will easily master the use of scales. Try the following:

- On a scale of 1:50, 40 mm represents: _____
- On a scale of 1:200, 70 mm represents:

  _____
- On a scale of 1:500, 40 mm represents:

  _____

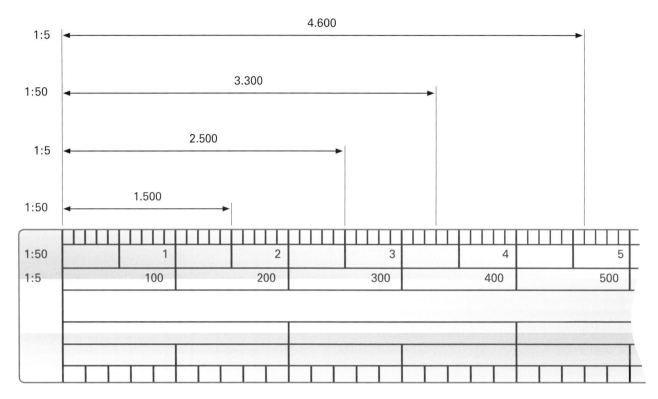

Figure 2.1 Rule with scales for maps and drawings

Scale drawings

Building plans are drawn to scale. Each length on the plan is in proportion (see Unit 1001, page 60) to the real length. On a drawing that has been drawn to a scale of 1 cm represents 10 m:

- a length of 5 cm represents an actual length of 5 × 10 = 50 m
- a length of 12 cm represents an actual length of 12 × 10 = 120 m
- an actual length of 34 m is represented by a line 34 ÷ 10 = 3.4 cm long.

Scales are often given as **ratios**. For example:

- a scale of 1:100 means that 1 cm on the drawing represents an actual length of 100 cm (or 1 m)
- a scale of 1:20 000 means that 1 cm on the drawing represents an actual length of 20 000 cm (or 20 m).

Table 2.2 shows some common scales used in the construction industry.

1:5	1 cm represents 5 cm	5 times smaller than actual size
1:10	1 cm represents 10 cm	10 times smaller than actual size
1:20	1 cm represents 20 cm	20 times smaller than actual size
1:50	1 cm represents 50 cm	50 times smaller than actual size
1:100	1 cm represents 100 cm = 1 m	100 times smaller than actual size
1:1250	1 cm represents 1250 cm = 12.5 m	1250 times smaller than actual size

Table 2.2 Common scales used in the construction industry

Now look at the following examples.

Example

A plan is drawn to a scale of 1:20. On the plan, a wall is 4.5 cm long. How long is the actual wall?

1 cm on the plan = actual length 20 cm

So 4.5 cm on the plan = actual length 4.5 × 20 = 90 cm or 0.9 m.

Example

A window is 3 m tall. How tall is it on the plan?

3 m = 300 cm

an actual length of 20 cm is 1 cm on the plan

an actual length of 5 × 20 = 100 cm is 5 × 1 cm on the plan

an actual length of 3 × 100 cm is 3 × 5 cm on the plan.

Therefore, the window is 15 cm tall on the plan.

Did you know?

To make scale drawings, architects use a scale rule. The different scales on the ruler give the equivalent actual length measurements for different lengths in cm, for each scale

Symbols and abbreviations

The use of symbols and abbreviations allows the maximum amount of information to be included on a drawing sheet in a clear way. Figure 2.2 shows some recommended drawing symbols for a range of building materials.

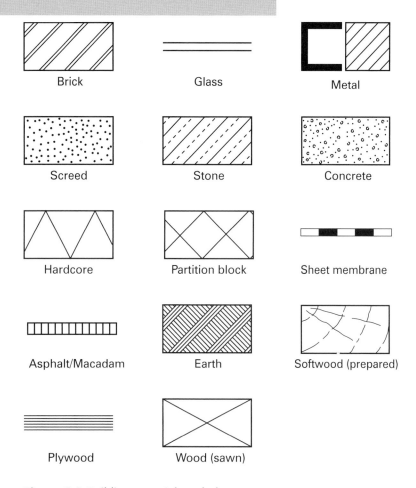

Figure 2.2 Building material symbols

Figure 2.3 illustrates the recommended methods for indicating different types of doors and windows and their direction of opening.

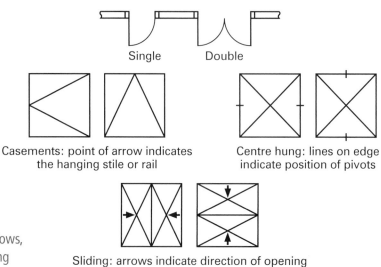

Figure 2.3 Doors and windows, type and direction of opening

Figure 2.4 shows some of the most frequently used graphical symbols, which are recommended in the British Standards Institute (see Unit 1001, page 52) standard BS 1192.

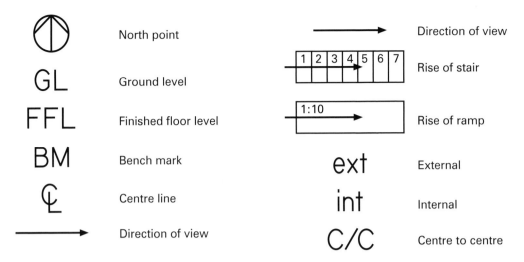

Figure 2.4 Graphical symbols used in the building industry

Table 2.3 lists some standard abbreviations used on drawings.

Item	Abbreviation	Item	Abbreviation
Airbrick	AB	Cement	ct
Asbestos	Asb	Column	col
Bitumen	Bit	Concrete	conc
Boarding	bdg	Cupboard	cpd
Brickwork	Bwk	Damp proof course	DPC
Building	Bldg	Damp proof membrane	DPM
Cast iron	CI	Drawing	dwg
Foundation	Fnd	Polyvinyl chloride	PVC
Hardboard	hdbd	Reinforced concrete	RC
Hardcore	hc	Satin anodised aluminium	SAA
Hardwood	Hwd	Satin chrome	SC
Insulation	insul	Softwood	swd
Joist	jst	Stainless steel	SS
Mild steel	MS	Tongue and groove	T&G
Plasterboard	pbd	Wrought iron	WI
Polyvinyl acetate	PVA		

Table 2.3 Standard abbreviations used on drawings

> **Remember**
>
> It is important to check that the documents you are working with are the most recent. If they have changed in any way, and you haven't been told, what you will be doing will be wrong

Location drawings, specifications and schedules

Specifications, drawings and schedules are the main reference documents for work on site and are used to plan all the work that takes place during the build. It is important that all these documents are accurate and correct, and that any changes made to them are clearly communicated to everyone working on the site.

Location drawings

Location drawings include block plans and site plans, and are used to show what the site will look like when it is completed. It is drawn to a chosen scale.

- **Block plans** – identify the proposed site by giving a bird's eye view of the site in relation to the surrounding area. An example is shown in Figure 2.5.
- **Site plans** – give the position of the proposed building and the general layout of the roads, services, drainage, etc. on site. An example is shown in Figure 2.6.

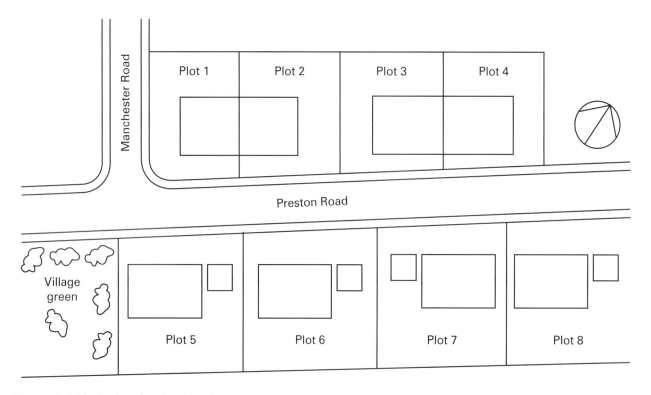

Figure 2.5 Block plan showing location

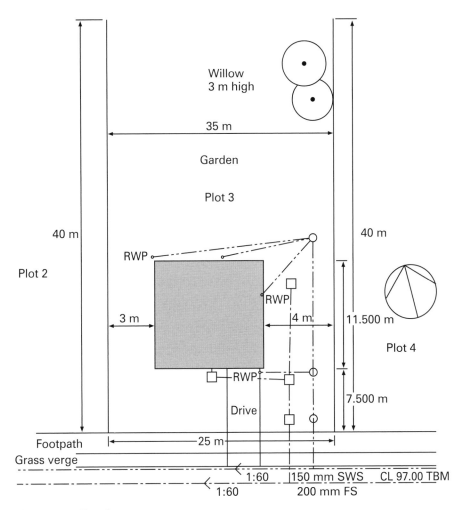

Figure 2.6 Site plan

Specifications

The specification or 'spec' is a document produced alongside the plans and drawings and is used to show information that can't be shown on the drawings. Specifications are almost always used, except in the case of very small contracts. A specification should contain:

- **site description** – a brief description of the site, including the address
- **restrictions** – what restrictions apply such as working hours or limited access
- **services** – what services are available, what services need to be connected and what type of connection should be used
- **materials description** – including type, size, quality, moisture content, etc.
- **workmanship** – including methods of fixing, quality of work and finish.

Unit 1002 Information, quantities and communicating with others

Figure 2.7 A good 'spec' helps avoid confusion when dealing with subcontractors or suppliers

The specification may also name subcontractors or suppliers, or give details such as how the site should be cleared, and so on.

Schedules

A schedule is used to record repeated design information that applies to a range of components or fittings. Schedules are mainly used on bigger sites where there are multiples of several types of house (four-bedroom, three-bedroom, three-bedroom with dormers, etc.), each type having different components and fittings. Schedules avoid the wrong component or fitting being put in the wrong house. Schedules can also be used on smaller jobs such as a block of flats with 200 windows, where there are six different types of window.

The need for a schedule depends on the complexity of the job and the number of repeated designs that there are. Schedules are mainly used to record repeated design information for:

- doors
- sanitary components
- windows
- heating components and radiators
- ironmongery
- kitchens
- joinery fitments.

A schedule is usually used with a range drawing and a floor plan.

Figures 2.8–2.10 show basic examples of these documents, using a window as an example:

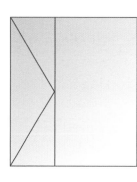

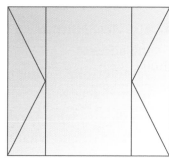

Window 1   Window 2   Window 3   Window 4   Window 5

Figure 2.8 Range drawing

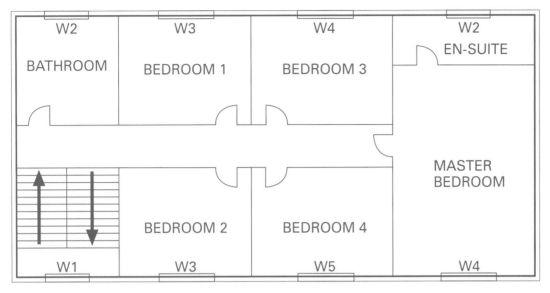

Figure 2.9 Floor plan

WINDOW SCHEDULE		
WINDOW	LOCATIONS	NOTES
Window 1	Stairwell	
Window 2	Bathroom En-suite	Obscure glass
Window 3	Bedroom 1 Bedroom 2	
Window 4	Bedroom 3 Master bedroom	
Window 5	Bedroom 4	

Figure 2.10 Schedule for the windows

The schedule in Figure 2.10 shows that there are five types of window, each differing in size and appearance; the range drawing shows what each type of window looks like; and the floor plan shows which window goes where. For example, the bathroom window is a type two window, which is 1200 × 600 × 50 cm with a top-opening sash and obscure glass.

Work programme

A work programme is a method of showing very easily what work is being carried out on a building and when. The most common form of work programme is a bar chart. Used by many site agents or supervisors, a bar chart lists the tasks that need to be done down the left side and shows a timeline across the top. A work programme is used to make sure that the relevant trade is on site at the correct time and that materials are delivered when needed. A site agent or supervisor can quickly tell from looking at the chart if work is keeping to schedule or falling behind.

Did you know?

The Gantt chart is named after Henry Gantt, an American engineer, who, in 1910, was the first to design and use this chart

Bar charts

The bar or Gantt chart is the most popular work programme. It is simple to construct and easy to understand. Bar charts have tasks listed in a vertical column on the left and a horizontal timescale running along the top.

Time in days										
Activity	1	2	3	4	5	6	7	8	9	10
Dig for foundation and service routes										
Lay foundations										
Run cabling, piping, etc. to meet existing services										
Build up to damp proof course										
Lay concrete floor										

Figure 2.11 Basic bar chart

Each task is given a proposed time, which is shaded in along the horizontal timescale. Timescales often overlap as one task often overlaps another.

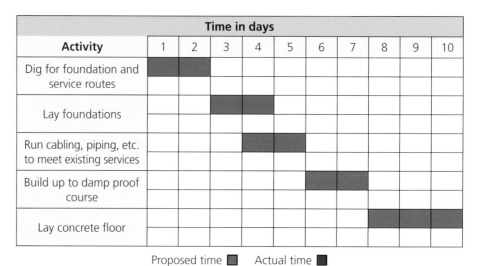

Time in days										
Activity	1	2	3	4	5	6	7	8	9	10
Dig for foundation and service routes	▓	▓								
Lay foundations			▓	▓						
Run cabling, piping, etc. to meet existing services				▓	▓					
Build up to damp proof course						▓	▓			
Lay concrete floor								▓	▓	▓

Proposed time ▓   Actual time ■

Figure 2.12 Bar chart showing proposed time for a contract

The bar chart can then be used to check progress. Often the actual time taken for a task is shaded in underneath the proposed time (in a different way or colour to avoid confusion). This shows how what *has* been done matches up to what *should* have been done.

A bar chart can therefore help you plan when to order materials or plant, see what trade is due in and when, and so on. A bar chart can also tell you if you are behind on a job; this information is vital if your contract contains a **penalty clause**.

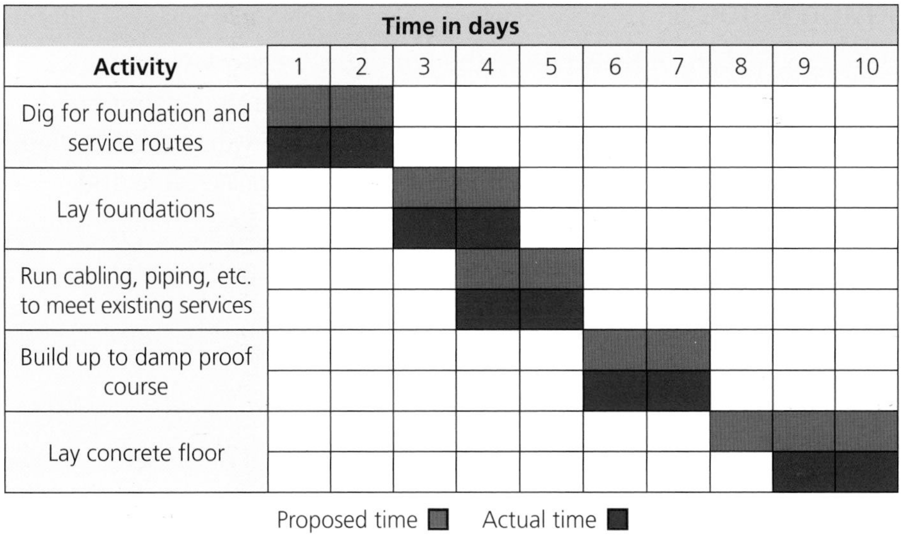
Time in days
Activity
1
2
3
4
5
6
7
8
9
10
Dig for foundation and service routes
Lay foundations
Run cabling, piping, etc. to meet existing services
Build up to damp proof course
Lay concrete floor
Proposed time
Actual time

Figure 2.13 Bar chart showing actual time halfway through a contract

When creating a bar chart, you should build in some extra time to allow for things such as bad weather, labour shortages, delivery problems or illness. It is also advisable to have contingency plans to help solve or avoid problems such as:

- capacity to work overtime to catch up time
- bonus scheme to increase productivity
- penalty clause on suppliers to try to avoid late or poor deliveries
- source of extra labour (for example from another site) if needed.

Good planning, with contingency plans in place, should allow a job to run smoothly and finish on time, leading to the contractor making a profit.

K2. Determining quantities of materials

The information contained in the drawings and specification for a project will tell you what materials you will need for the job. You will use this information to determine the quantity of each type of material you will need. To work this out you need to know the methods used to calculate basic **estimates** of material quantity.

Key term

Penalty clause – a condition written into the contract that states the work must be completed to the required quality by a certain date. If the job over-runs, the contractor will not be paid the full amount for each day the job runs over

Did you know?

Bad weather is the main external factor responsible for delays on building sites in the UK. A Met Office survey showed that UK construction companies experience problems caused by the weather on average 26 times a year

Key term

Estimate – to assess something such as a job to be done, and to state a likely price for it

When making calculations there are several resources you will find useful. These include:

- diagrams and plans
- calculators
- conversion tables
- scale rules.

Functional skills

FM 1.2.1f relates to using whole numbers. These may be positive or negative numbers. Think of a situation where you would use these numbers – e.g. attendance figures at a football match or the temperature in both winter and summer.

Numbers

Place value

0, 1, 2, 3, 4, 5, 6, 7, 8 and 9 are the ten digits we work with. We can write any number you can think of, however huge, using any combination of these ten digits. In a number, the value of each digit depends on its 'place value'. Table 2.4 is a place value table and shows how the value of digit 2 is different, depending on its position.

Millions	Hundred thousands	Ten thousands	Thousands	Hundreds	Tens	Units	Value
2	9	4	1	3	7	8	2 million
	2	5	3	1	0	7	2 hundred thousand
	7	**2**	5	6	6	4	2 × ten thousand = 20 thousand
		6	**2**	4	9	2	2 thousands
		5	6	**2**	9	1	2 hundreds
			8	4	**2**	7	2 tens = 20
				1	6	**2**	2 units

Table 2.4 Place value table for the digit 2

Positive numbers

A positive number is a number that is greater than zero. If we make a number line, positive numbers are all the numbers to the right of zero.

0 1 2 3 4 5 6 7 8 9 10 11 12 13…

⟶

Positive numbers

Negative numbers

A negative number is a number that is less than zero. If we make another number line, negative numbers are all the numbers to the left of zero.

$$...-13 \ -12 \ -11 \ -10 \ -9 \ -8 \ -7 \ -6 \ -5 \ -4 \ -3 \ -2 \ -1 \ 0$$

Negative numbers

Making calculations

There are several calculation methods that are used to calculate the area of basic shapes. The main ones you will need to use are:

- addition
- subtraction
- multiplication
- division.

These methods are used to calculate the basic areas of a series of shapes that you will encounter on plans and diagrams.

Functional skills

FM 1.2.1a relates to mathematical procedures such as addition and subtraction. You will need to organise numbers in units, tens, hundreds and thousands in order to carry out the task.

Addition

When adding numbers using a written method, write digits with the same place value in the same column. For example, to work out what 26 + 896 + 1213 is, write the calculation:

Add up the digits in columns, starting with the units column.

```
      2 6
      8 9 6
    1 2 1 3
    -------
    2 1 3 5
    1 1 1
```

(4) 1 + 1 = 2

(1) 6 + 6 + 3 = 15
Write 5 in the digits column and carry 1 ten to the tens column.

(3) 8 + 2 = 10
Add the carried 1 from the tens column 10 + 1 = 11. Write 1 and carry 1 to the thousands column.

(2) 2 + 9 + 1 = 12
Add the carried 1 from the digits column 12 + 1 = 13. Write 3 in the tens column and carry the 1 hundred to the hundreds column.

To add numbers with decimals, write the numbers with the decimal points in line:

$$
\begin{array}{r}
4.56 \\
10.2 \\
\underline{0.32} \\
\underline{15.08}
\end{array}
$$

In a problem, the following words mean you need to add.

- **Total** – What is the total of 43 and 2457? (43 + 2457)

- **Sum** – What is the sum of 56 and 345? (56 + 345)

- **Increase** – Increase 3467 by 521 (3467 + 521)

Subtraction

When subtracting numbers using a written method, write digits with the same place value in the same column.

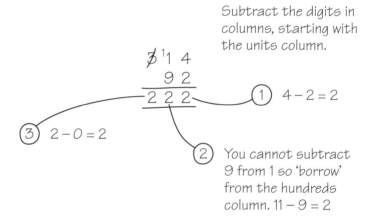

Subtract the digits in columns, starting with the units column.

① 4 − 2 = 2

② You cannot subtract 9 from 1 so 'borrow' from the hundreds column. 11 − 9 = 2

③ 2 − 0 = 2

In a problem, the following words and phrases mean you need to subtract.

- **Find the difference** – Find the difference between 200 and 45 (200 − 45)
- **Decrease** – Decrease 64 by 9 (64 − 9)
- **How much greater than** – How much greater than 98 is 110? (110 − 98)

Multiplication

Knowing multiplication tables up to 10 × 10 helps with multiplying single digit numbers. You can use multiplication facts you already know to work out other multiplication calculations.

Example

What is 20 × 12?

You know that 20 = 2 × 10

So, 20 × 12 = 2 × 10 × 12

    = 2 × 12 × 10

    = 24 × 10 = 240

To multiply larger numbers you can write the calculation in columns or use the grid method. Both methods work by splitting the calculation into smaller ones.

Multiplying using columns

```
      2 5
 ×    3 6
    1 5 0        6 × 25 ⎫ Add these to
    7 5 0       30 × 25 ⎭ find 36 × 25
    9 0 0
```

Multiplying using the grid method

36 × 25 =

×	20	5
30	600	150
6	120	30

```
30 × 20 = 600
30 ×  5 = 150
 6 × 20 = 120
 6 ×  5 =  30
          900
```

Division

Division is the opposite of multiplication. Knowing the multiplication tables up to 10 × 10 also helps with division. Each multiplication fact gives two related division facts. For example:

4 × 6 = 24   24 ÷ 6 = 4   24 ÷ 4 = 6

Short division

When dividing by a single digit number, use short division.

161 √ 7 =

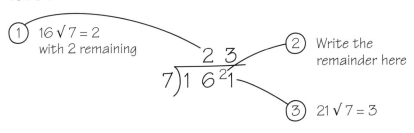

① 16 √ 7 = 2 with 2 remaining

② Write the remainder here

③ 21 √ 7 = 3

```
    2 3
7)1 6²1
```

Long division

When dividing by 10 or more, use long division.

$$12 \overline{) 2952} = 246$$

$2 \times 12 = 24$

$29 - 24 = 5$. Bring down the next 5

$12 \times 4 = 48$

$55 - 48 = 7$. Bring down the 2

$12 \times 6 = 72$

Estimating

Did you know?

You can use values rounded to 1 s.f. to estimate approximate areas and prices

Sometimes an accurate answer to a calculation is not required. You can estimate an approximate answer by rounding all the values in the calculation to the nearest whole number. For example:

Estimate the answer to the calculation 4.9×3.1

4.9 rounds to 5

3.1 rounds to 3

A sensible estimate is $5 \times 3 = 15$

Measures

Functional skills

Estimation is a functional skill used in calculations and is part of the **FM** 1.2.1a. This skill is crucial when measuring and working out amounts of materials. As with all the mathematical procedures, practise in these sections will help you understand functional skills.

In brickwork, quantities of material are presented in measurements. These measurements are used when ordering, and in plans and specifications. The mathematic skills described above will enable you to use these units of measurement.

Units of measurement

The metric units of measurement are shown in Table 2.5.

Length	millimetres (mm) centimetres (cm) metres (m) kilometres (km)
Mass (weight)	grams (g) kilograms (kg) tonnes (t)
Capacity (the amount a container holds)	millilitres (ml) centilitres (cl) litres (l)

Table 2.5 Units of measurement

Remember

Metric units are all based on 10, 100, 1000 – which makes it easy to convert between units

milli means one thousandth	$1 \text{ mm} = \frac{1}{1000} \text{ m}$	$1 \text{ ml} = \frac{1}{1000} \text{ litre}$
centi means one hundredth	$1 \text{ cm} = \frac{1}{100} \text{ m}$	$1 \text{ cl} = \frac{1}{1000} \text{ litre}$
kilo means one thousand	$1 \text{ kg} = 1000 \text{ g}$	$1 \text{ km} = 1000 \text{ m}$

Table 2.6 shows some useful metric conversions.

Length	Mass	Capacity
1 cm = 10 mm	1 kg = 1000 g	1 l = 100 cl = 1000 ml
1 m = 100 cm = 1000 mm	1 tonne = 1000 kg	
1 km = 1000 m		

Table 2.6 Useful metric conversions

Remember

To convert from a smaller unit to a larger one – divide
To convert from a larger unit to a smaller one – multiply

To convert 2657 mm to metres:	2657 ÷ 1000 = 2.657 m
To convert 0.75 tonnes to kg:	0.75 × 1000 = 750 kg

For calculations involving measurements, you need to convert all the measurements into the same unit.

Example

A plasterer measures the lengths of cornice required for a room. He writes down the measurements as 175 cm, 2 m, 225 cm, 1.5 m. To work out the total length of cornice needed, we first need to write all the lengths in the same unit:

175 cm

2 m = 2 × 100 = 200 cm

1.5 m = 1.5 × 100 = 150 cm

225 cm

So the total length is:

175 + 200 + 225 + 150 = 750 cm

Imperial units

In the UK we still use some imperial units of measurement (see Table 2.7).

Length	inches feet yards miles
Mass (weight)	ounces pounds stones
Capacity (the amount a container holds)	pints gallons

Table 2.7 Some imperial units of measurement

To convert from imperial to metric units, use the approximate conversions shown in Table 2.8.

Did you know?

A useful rhyme to help you remember the pints to litre conversion is: 'a litre of water is a pint and three quarters'

Length	Mass	Capacity
1 inch = 2.5 cm	2.2 pounds = 1 kg	1.75 pints = 1 litre
1 foot = 30 cm	1 ounce = 25 g	1 gallon = 4.5 litres
5 miles = 8 km		

Table 2.8 Converting imperial measurements to metric

Functional skills

FM 1.2.1c, d and l relate to using formula to calculate perimeters and areas. This information is important in understanding the amount of a material required to carry out a building task.

Example

If a wall is 32 feet long, what is its approximate length in metres?

   1 foot = 30 cm

So 32 feet = 32 × 30 cm = 960 cm = 9.6 m.

Calculating perimeters and areas

Perimeter of shapes with straight sides

The perimeter of a shape is the distance all around the outside of the shape. To find the perimeter of a shape, measure all the sides and then add the lengths together.

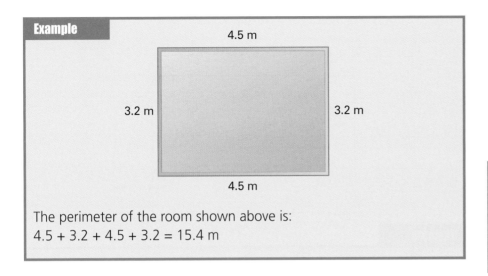

The perimeter of the room shown above is:
4.5 + 3.2 + 4.5 + 3.2 = 15.4 m

Units of area

The area of a two-dimensional (2-D; flat) shape is the amount of space it covers. Area is measured in square units, e.g. square centimetres (cm^2) and square metres (m^2).

The area of the square below is $10 \times 10 = 100$ mm^2 or 1 cm^2.

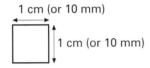

The area of the square below is $100 \times 100 = 10\ 000$ cm^2 or 1 m^2.

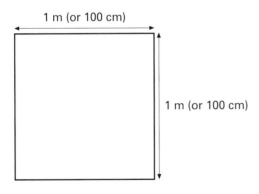

Remember

$1\ cm^2 = 100\ mm^2$

$1\ m^2 = 10\ 000\ cm^2$

Area of shapes with straight sides

The rectangle below is drawn on squared paper. Each square has an area 1 cm^2.

You can find the area by counting the squares:

  Area = 6 squares = 6 cm^2

You can also calculate the area by multiplying the number of squares in a row by the number of rows:

  $3 \times 2 = 6$

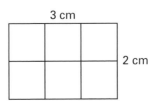

The area of a rectangle with length l and width w is:

$$A = l \times w$$

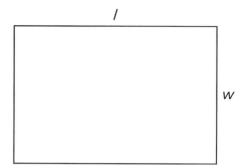

Example

If the length of a rectangular room is 3.6 m and the width is 2.7 m, the area is:

$$A = 3.6 \times 2.7 = 9.72 \text{ m}^2$$

Area of a triangle

The area of a triangle is given by the formula:

$$A = \tfrac{1}{2} \times h \times b$$

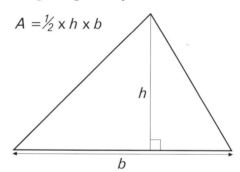

Key term

Perpendicular – at right angles to

$A = \frac{1}{2} \times h \times b$ where h is the **perpendicular** height and b is the length of the base. The perpendicular height is drawn to meet the base at right angles (90°).

Example

What is (a) the area and (b) the perimeter of the triangle below?

1. Perpendicular height = 4 m, base = 3 m

$$A = \frac{1}{2} \times h \times b = \frac{1}{2} \times 4 \times 3 = 6 \text{ m}^2$$

2. Perimeter = 5 + 4 + 3 = 12 m

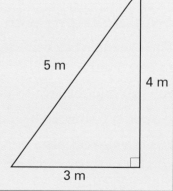

Pythagoras' theorem

You can use Pythagoras' theorem to find unknown lengths in right-angled triangles. In a right-angled triangle:

- one angle is 90° (a right angle)
- the longest side is opposite the right angle and is called the **hypotenuse**.

Pythagoras' theorem says that for any right-angled triangle with sides a and b and hypotenuse c,

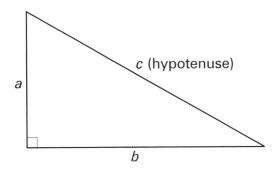

$$c^2 = a^2 + b^2$$

Example

What is the length of the hypotenuse of the triangle below?

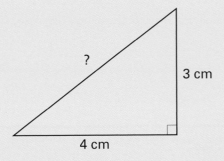

We know $c^2 = a^2 + b^2$

where c is the hypotenuse.

$c^2 = a^2 + b^2$

$\quad = 9 + 16$

$\quad = 25$

$c\ = \sqrt{25} = 5$

Therefore the hypotenuse is 5 cm.

a^2 is 'a squared' and is equal to $a \times a$. a is '3 squared' and is equal to 3×3. The opposite or inverse of squaring is finding the square root. $\sqrt{25}$ means 'the square root of 25': $5 \times 5 = 25$, so $\sqrt{25} = 5$.

Learning these squares and square roots will help with Pythagoras' theorem calculations. Table 2.9 shows some squares and square roots you will often find useful to know.

$1^2 = 1 \times 1 = 1$	$\sqrt{1} = 1$
$2^2 = 2 \times 2 = 4$	$\sqrt{4} = 2$
$3^2 = 3 \times 3 = 9$	$\sqrt{9} = 3$
$4^2 = 4 \times 4 = 16$	$\sqrt{16} = 4$
$5^2 = 5 \times 5 = 25$	$\sqrt{25} = 5$
$6^2 = 6 \times 6 = 36$	$\sqrt{36} = 6$
$7^2 = 7 \times 7 = 49$	$\sqrt{49} = 7$
$8^2 = 8 \times 8 = 64$	$\sqrt{64} = 8$
$9^2 = 9 \times 9 = 81$	$\sqrt{81} = 9$
$10^2 = 10 \times 10 = 100$	$\sqrt{100} = 10$

Table 2.9 Useful squares and square roots

Using Pythagoras' theorem to find the shorter side of a triangle

You can rearrange Pythagoras' theorem like this:

$$c^2 = a^2 + b^2$$

$$a^2 = c^2 - b^2$$

Example

What is the length of side a in the right-angled triangle below?

$a^2 = c^2 - b^2$

$\quad = 12^2 - 6^2$

$\quad = 144 - 36 = 108$

$a \quad = \sqrt{108} = 10.3923...$ (using the $\sqrt{}$ key on a calculator)

$\quad = 10.4$ cm (to 1 decimal place (d.p.)).

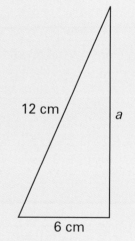

12 cm

a

6 cm

Example

You can also use Pythagoras' theorem to find the perpendicular height of a triangle. For example, if we wanted to find the area of the triangle below, we would need to find the perpendicular height:

Using Pythagoras' theorem:

$h^2 = 6^2 - 3^2$

$= 36 - 9 = 27$

$h = \sqrt{27} = 5.196... = 5.2$ cm (to 1 d.p.)

Area $= \frac{1}{2} \times b \times h$

$= \frac{1}{2} \times 5 \times 5.2$ Base length $= 3 + 2$ cm

$= 13$ cm^2

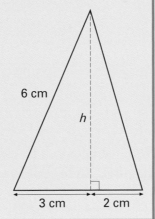

6 cm

h

3 cm   2 cm

Areas of composite shapes

Composite shapes are made up of simple shapes such as rectangles and squares. To find the area, divide up the shape and find the area of each part separately. For example, to work out the area of the L-shaped room below:

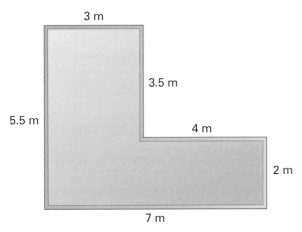

3 m

3.5 m

5.5 m

4 m

2 m

7 m

First divide it into two rectangles, A and B:

Area of rectangle A = 3 × 5.5 = 16.5 m^2

Area of rectangle B = 4 × 2 = 8 m^2

Total area of room = 16.5 + 8 = 24.5 m^2

You could also divide the rectangle in the example above into two different rectangles, C and D, like this:

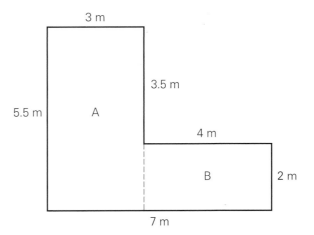

3 m

3.5 m

5.5 m   A

4 m

B   2 m

7 m

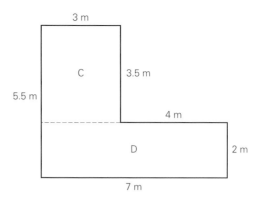

Check that you get the same total area.

Some shapes can be divided into rectangles and triangles. For example, to find the area of the wooden floor below:

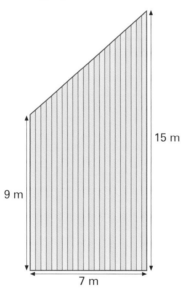

Divide the floor into a right-angled triangle A and a rectangle B

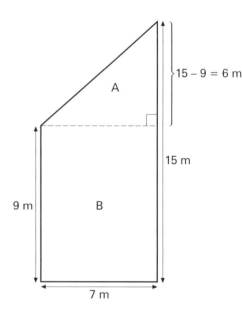

Triangle A has a vertical height of 6 m and a base of 7 m

Area of the triangle A $= \frac{1}{2} \times b \times h$

$\qquad = \frac{1}{2} \times 7 \times 6 = 21$ m^2

Area of the rectangle B $= 9 \times 7 = 63$ m^2

Total area $= 21 + 63 = 84$ m^2.

Circumference of a circle

The formula for the **circumference** of a circle of **radius** r is:

$\qquad C = 2\pi r$

Key terms

Circumference – a circle's perimeter, i.e. the distance all the way around the outside

Radius – the distance from the centre of a circle to its outside edge

The diameter is the distance across the circle through the centre (diameter = 2 × radius).

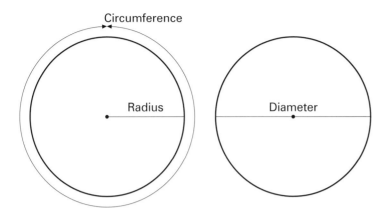

Circumference

Radius

Diameter

$\pi = 3.141\ 592\ 654\ldots$

To estimate the circumference of a circle, use $\pi = 3$. For more accurate calculations use $\pi = 3.14$, or the π key on a calculator.

Example

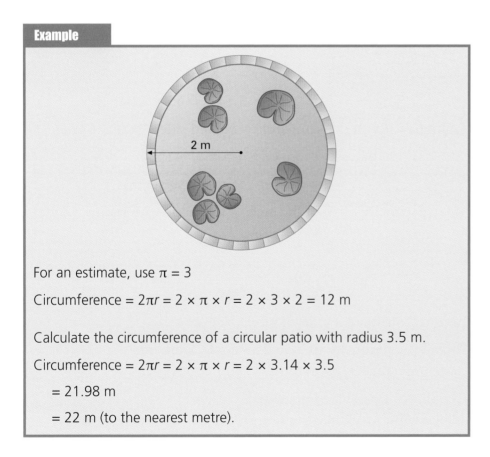

2 m

For an estimate, use $\pi = 3$

Circumference = $2\pi r = 2 \times \pi \times r = 2 \times 3 \times 2 = 12$ m

Calculate the circumference of a circular patio with radius 3.5 m.

Circumference = $2\pi r = 2 \times \pi \times r = 2 \times 3.14 \times 3.5$

= 21.98 m

= 22 m (to the nearest metre).

Area of a circle

The formula for the area of a circle of radius r is:

Area = πr^2

Did you know?

If you are given the diameter of the circle, you need to halve the diameter to find the radius

We can calculate the area of a circle with radius 3.25 m as:

Area = πr^2

$= \pi \times r^2$

$= 3.14 \times 3.25 \times 3.25 = 33.166\ 25$

$= 33\ m^2$ (to the nearest metre).

Area and circumference of part circles and composite shapes

You can use the formulae for circumference and area of a circle to calculate perimeters and areas of parts of circles, and shapes made from parts of circles. For example, we can work out the perimeter and area of the semicircular window below.

The diameter of the semicircle is 1.3 m, so the radius is $1.3 \div 2 = 0.65$ m.

The length of the curved side is half the circumference of the circle with radius 0.65 m.

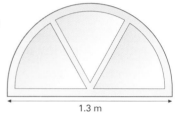
1.3 m

Length of curved side $= \frac{1}{2} \times 2\pi r = \frac{1}{2} \times 2 \times \pi \times r$

$= \frac{1}{2} \times 2 \times 3.14 \times 0.65 = 2.041$ m

Circumference of the semicircle = curved side + straight side

$= 2.041 + 1.3 = 3.341$m

$= 3.34$ m (to the nearest cm)

Area of semicircle = half the area of the circle with radius 0.65m

$= \frac{1}{2} \times \pi r^2 = \frac{1}{2} \times \pi \times r^2$

$= \frac{1}{2} \times 3.14 \times 0.65 \times 0.65 = 0.663\ 325\ m^2$

$= 0.66\ m^2$ (to 2 d.p.)

To find the area of a quarter circle, use $\frac{1}{4}\pi r^2$.

To find the perimeter of a quarter circle, work out $\frac{1}{4}$ circumference + 2 × radius.

To find the area of a composite shape including parts of circles, divide it into circles and simple shapes and find the areas separately.

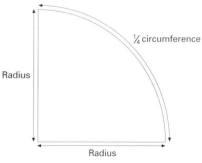
¼ circumference

Radius

Radius

K3. Relaying information in the workplace

Communication, in the simplest of terms, is a way or means of passing on information from one person to another. Communication is very important in all areas of life and we often do it without even thinking about it. You will need to communicate well when you are at work, no matter what job you do. What would happen if someone couldn't understand something you had written or said? If we don't communicate well, how will other people know what we want or need and how will we know what other people want?

Companies that do not establish good methods of communicating with their workforce, or with other companies, will not function properly. They will also end up with bad working relationships. Co-operation and good communication is vital to achieve good working relationships.

Message taking

One of the most common reasons for communicating is to give a message to someone. You could often be the 'channel' passing a message from one person to another (the 'third party').

To ensure good and efficient communication there are certain things you can do when taking messages. When a phone call is made, many things may be discussed. Some of these things may need to be passed on to a third party. The main things that you need to record when receiving a phone call are:

- date
- time
- contact name and details.

The topic of conversation can also be recorded but, if it's complex or lengthy, this may not be possible. The most important thing to take is the person's contact details. This will allow the person the message is for to get in direct contact with the person sending the message, so that further conversations can take place.

Taking accurate information from a conversation is important as other people may use this as a source of information for making decisions. For example, a person receiving a message may decide when to call back, based on a message. If the message didn't indicate the matter was urgent, the person may delay returning the call, which could have serious consequences.

Unit 1002

Some companies have their own forms to cover such things as scaffolding checks

While working through this unit, you will be practising the functional skills **FE** 1.2.1 – 1.2.3. These relate to reading and understanding information. As with the other sections in the book, if there are any words or phrases you do not understand, use a dictionary, look them up using the internet or discuss with your tutor.

Information relevant to communication

There are a number of different sources of information on site that are important for effective communication. By using these sources of information, you can improve the effectiveness of your communication.

General site paperwork

No building site could function properly without a certain amount of paperwork. Here is a brief description of some of the documents you may encounter, besides the ones already mentioned in the first part of this unit.

Timesheets

Timesheets record hours worked, and are completed by every employee individually. Some timesheets are basic, asking just for a brief description of the work done each hour, but some can be more detailed. In some cases timesheets may be used to work out how many hours the client will be charged for.

P. Gresford Building Contractors

Timesheet _____

Employee _____   **Project/site** _____

Date	Job no.	Start time	Finish time	Total time	Travel time	Expenses
M						
Tu						
W						
Th						
F						
Sa						
Su						
Totals						

Employee's signature _____

Supervisor's signature _____

Date _____

Figure 2.14 Timesheet

Day worksheets

Day worksheets are often confused with timesheets. However, they are different as day worksheets are used when there is no price or estimate for the work and so enable the contractor to charge for the work. Day worksheets record work done, hours worked and, sometimes, the materials used.

Job sheets

Job sheets are similar to day worksheets – they record work done – but are used when the work has already been priced. Job sheets enable the worker to see what needs to be done and the site agent or working foreman to see what has been completed.

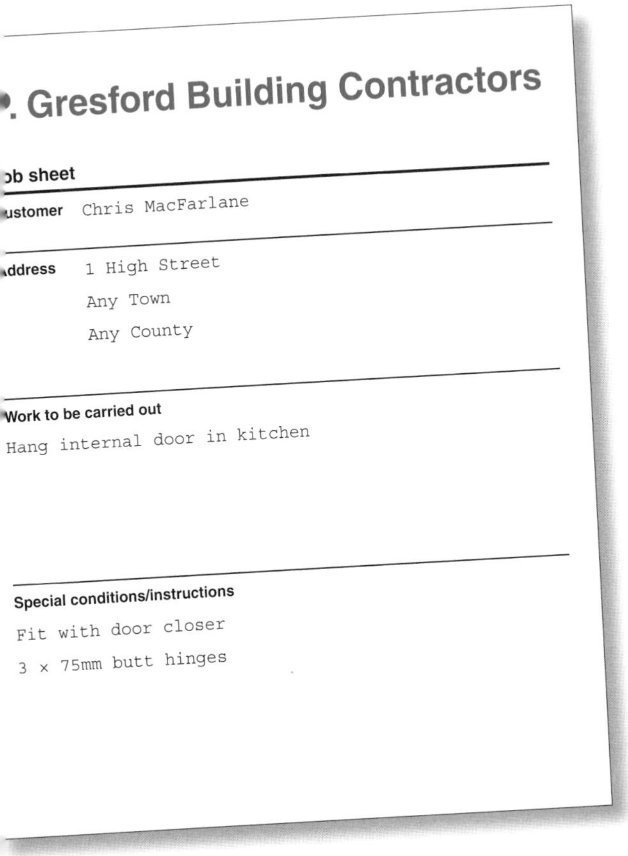

P. Gresford Building Contractors

Job sheet

Customer   Chris MacFarlane

Address   1 High Street

Any Town

Any County

Work to be carried out

Hang internal door in kitchen

Special conditions/instructions

Fit with door closer

3 x 75mm butt hinges

Figure 2.15 Day worksheet

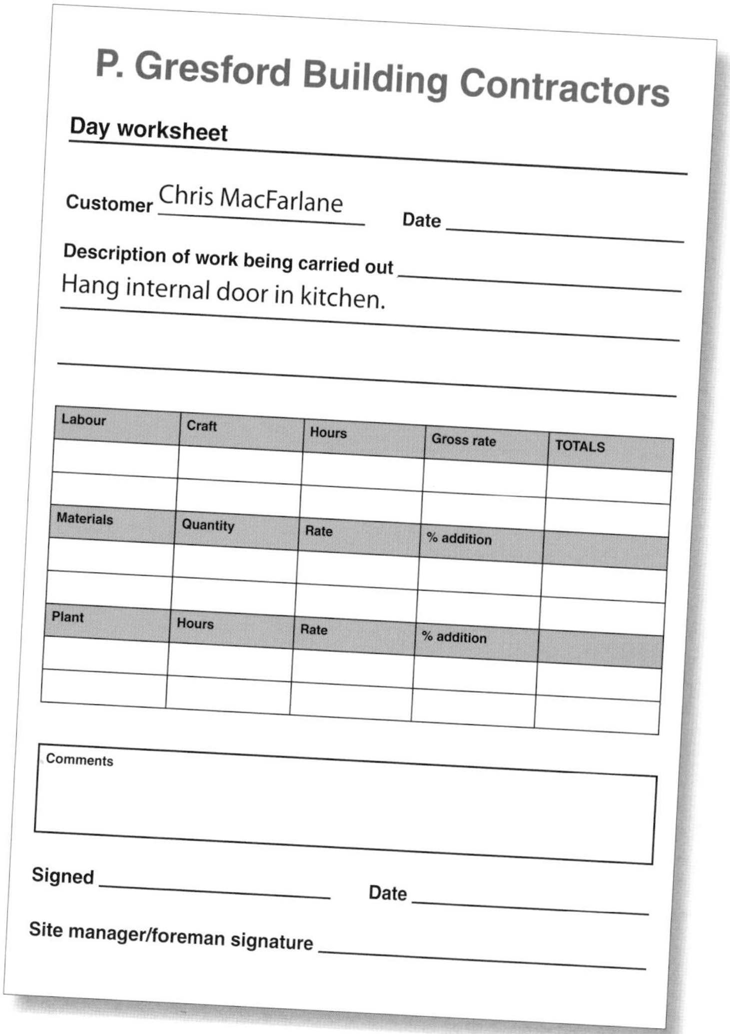

P. Gresford Building Contractors

Day worksheet

Customer Chris MacFarlane   Date _____

Description of work being carried out _____

Hang internal door in kitchen.

Labour	Craft	Hours	Gross rate	TOTALS
Materials	Quantity	Rate	% addition	
Plant	Hours	Rate	% addition	

Comments

Signed _____   Date _____

Site manager/foreman signature _____

Figure 2.16 Job sheet

VARIATION TO PROPOSED WORKS AT 123 A STREET

REFERENCE NO:

DATE _____

FROM _____

TO _____

POSSIBLE VARIATIONS TO WORK AT 123 A STREET

ADDITIONS

OMISSIONS

SIGNED -

Variation order

This sheet is used by the architect to make any changes to the original plans, including omissions, alterations and extra work.

Figure 2.17 Variation order

CONFIRMATION FOR VARIATION TO PROPOSED WORKS AT 123 A STREET

REFERENCE NO:

DATE _____

FROM _____

TO _____

I CONFIRM THAT I HAVE RECEIVED WRITTEN INSTRUCTIONS
FROM _____
POSITION _____
TO CARRY OUT THE FOLLOWING POSSIBLE VARIATIONS TO THE ABOVE NAMED CONTRACT

ADDITIONS

OMISSIONS

SIGNED -

Confirmation notice

This is a sheet given to the contractor to confirm any changes that have been made in the variation order, so that the contractor can go ahead and carry out the work.

Figure 2.18 Confirmation notice

Orders/Requisitions

A requisition form or order is used to order materials or components from a supplier.

P. Gresford Building Contractors

Requisition form

Supplier _____   Order no. _____
_____   Serial no. _____

Tel no. _____   Contact _____

Fax no. _____   Our ref _____

Contract/Delivery address/Invoice address   Statements/applications
_____   for payments to be sent to
_____   _____

Tel no. _____   _____

Fax no. _____   _____

Item no.	Quantity	Unit	Description	Unit price	Amount

Total £ _____

Payment terms _____   Date
_____

Originated by
_____

Authorised by
_____

Figure 2.19 Requisition form

Bailey & Sons Ltd

Building materials supplier

Tel: 01234 567890

Your ref: AB00671

Our ref: CT020

Order no: 67440387

Date: 17 Jul 2010

Invoice address:
Carillion Training Centre,
Depford Terrace, Sunderland

Delivery address:
Same as invoice

Description of goods	Quantity	Catalogue no.
OPC 25kg	10	OPC1.1

Comments:

Date and time of receiving goods:

Name of recipient (caps):

Signature:

Figure 2.20 Delivery note

Remember

Remember

If there are any discrepancies or problems with a delivery such as poor quality or damaged goods, you should write on the delivery note what is wrong *before* you sign it. You should also make sure that the site agent is informed so that they can rectify the problem

Remember

Invoices may need paying by a certain date – fines for late payment can sometimes be incurred. Therefore, it is important that they are passed on to the finance office or financial controller promptly

Delivery notes

Delivery notes are given to the contractor by the supplier, and list all the materials and components being delivered. Each delivery note should be checked for accuracy against the order (to ensure what is being delivered is what was asked for) and against the delivery itself (to make sure that the delivery matches the delivery note).

Invoices

Invoices come from a variety of sources such as suppliers or subcontractors, and state what has been provided and how much the contractor will be charged for it.

INVOICE

L Weeks Builders
4th Grove
Thomastown

JARVIS BUILDING SUPPLIES
3rd AVENUE
THOMASTOWN

Quantity	Description	Unit price	Vat rate	Total
30	Galvanised joint hangers	£1.32	17.5%	£46.53
			TOTAL	£46.53

To be paid within 30 days from receipt of this invoice

Please direct any queries to 01234 56789

Figure 2.21 Invoice

Delivery records

Delivery records list all deliveries over a certain period (usually a month), and are sent to the contractor's head office so that payments can be made.

Daily report/site diary

This is used to pass general information (deliveries, attendance, etc.) to a company's head office.

JARVIS BUILDING SUPPLIES
3RD AVENUE
THOMASTOWN

Customer ref_____

Customer order date_____

Delivery date_____

Item no	Qty Supplied	Qty to follow	Description	Unit price
1	30	0	Galvanised joinst hangers	£1.32

Delivered to: L Weeks builders
4th Grove
Thomastown

Customer signature _ _ _ _ _ _ _ _ _ _ _ _ _

Figure 2.22 Delivery record

DAILY REPORT/SITE DIARY

PROJECT_____
DATE_____

Identify any of the following factors, which are affecting or may affect the daily work activities and give a brief description in the box provided

WEATHER ( )   ACCESS ( )   ACCIDENTS ( )   SERVICES ( )
DELIVERIES ( )   SUPPLIES ( )   LABOUR ( )   OTHER ( )

SIGNED _ _ _ _ _ _ _ _ _ _ _ _ _
POSITION _ _ _ _ _ _ _ _ _ _ _ _ _

Figure 2.23 Daily report or site diary

Policies and procedures

Most companies will have their own policies and procedures in the workplace. All employees will be expected to follow these.

A policy states what the company expects to be done in a certain situation. Companies usually have policies for most things ranging from health and safety to materials orders. For example, in a health and safety policy the company will expect all employees to abide by rules and regulations and be safe.

To ensure that the policies are followed, the company will use certain procedures for working. For example, using a certain form to record data or using a certain method of working.

These policies and procedures are vital in a large organisation that may be doing work in several different sites in different locations. A senior manager should be able to walk into any site at any location and see exactly the same set up with the same forms and procedures being used everywhere.

Site rules

Site rules will cover most things ranging from safety to security. The company will have general policies covering rules that everyone on site must follow. These include important but basic things such as hours of work and behaviour.

However, a local site may have additional rules that apply only to that site. This is because each site will have some different situations. Site rules deal with those situations that could occur *only* on that site.

Remember

Your company's rules should be explained to you when you first start work. Any additional site rules should be made clear at your site induction (see Unit 1001, page 12)

Positive and negative communication

You can communicate in a variety of ways, and the main methods of communication are explained below, with the advantages and disadvantages of each method.

The key point to remember is to make all your communication positive. Positive communication will basically have a positive outcome with the message being communicated successfully. This will lead to things getting done right the first time. Negative communication will have the opposite effect and may lead to costly delays.

For positive communication, you need to ensure that no matter what method you use, the communication is clear, simple and – importantly – communicated to the right people.

Methods of communication

There are many different ways of communicating with others and they all generally fit into one of these four categories:

- verbal communication (speaking), for example talking face to face, over the telephone or by radio
- written communication, for example sending a letter or a memo
- body language, for example the way we stand or our facial expressions
- electronic communication, for example e-mail, **fax** and text messages.

Each method of communicating has some good points (advantages) and some bad points (disadvantages).

> **Key term**
>
> **Fax** – short for facsimile, which is a kind of photocopy that can be sent by dialling a phone number on a fax machine

Verbal communication

Verbal communication is the most common method we use to communicate with each other. If two people don't speak the same language or if someone speaks very quietly or not very clearly, verbal communication cannot be effective. Working in the construction industry you may communicate verbally with other people face to face, over the telephone or by radio/walkie-talkie.

Advantages	Disadvantages
Verbal communication is instant, easy and can be repeated or rephrased until the message is understood.	Verbal communication can be easily forgotten as there is no physical evidence of the message. Because of this it can be easily changed if passed to other people. A different accent or use of slang language can sometimes make it difficult to understand what a person is saying.

Written communication

Written communication can take the form of letters, messages, notes, instruction leaflets and drawings among others.

Advantages	Disadvantages
There is physical evidence of the communication and the message can be passed on to another person without it being changed. It can also be read again if it is not understood.	Written communication takes longer to arrive and understand than verbal communication and body language. It can also be misunderstood or lost. If it is handwritten, the reader may not be able to read the writing if it is messy.

Messages

To: Andy Rodgers

Date: Tues 10 Nov   **Time:** 11.10

Message: Mark from Stokes called with a query about the recent order. Please phone asap. (tel 01234 567 890)

Message taken by: Lee Barber

Figure 2.24 A message is a form of written communication

Body language

It is said that, when we are talking to someone face to face, only 10 per cent of our communication is verbal. The rest of the communication is body language and facial expression. This form of communication can be as simple as the shaking of a head from left to right to mean 'no' or as complex as the way someone's face changes when they are happy or sad or the signs given in body language when someone is lying.

We often use hand gestures as well as words to get across what we are saying, to emphasise a point or give a direction.

Some people communicate entirely through a form of body language called sign language.

Advantages	Disadvantages
If you are aware of your own body language and know how to use it effectively, you can add extra meaning to what you say. For example, when you are talking to a client or a colleague, even if the words you are using are friendly and polite and if your body language is negative or unfriendly, the message that you are giving out could be misunderstood. By simply maintaining eye contact, smiling and not folding your arms, you have made sure that the person you are communicating with has not got a mixed or confusing message. Body language is quick and effective. A wave from a distance can pass on a greeting without being close, and using hand signals to direct a lorry or a load from a crane is instant and doesn't require any equipment such as radios.	Some gestures can be misunderstood, especially if they are given from very far away. Also, some gestures that have one meaning in one country or culture can have a completely different meaning in another.

Figure 2.25 Try to be aware of your body language

Functional skills

At the end of this unit you will have the opportunity to answer a series of questions on the material you have learnt. By answering these questions you will be practising the following functional skills:

FE 1.2.3 Read different texts and take appropriate action.

FE 1.3.1 – 1.3.5 Write clearly with a level of detail to suit the purpose.

FM 1.1.1 Identify and select mathematical procedures.

FM 1.2.1c Draw shapes.

Electronic communication

Electronic communication is becoming more and more common and easy with the advances in technology. Electronic communication can take many forms such as text messages, e-mail and fax. It is now even possible to send and receive e-mails via a mobile phone, which allows important information to be sent or received from almost anywhere in the world.

Advantages	Disadvantages
Electronic communication takes the best parts from verbal and written communication in as much as it is instant, easy and there is a record of the communication being sent. Electronic communication goes even further as it can tell the sender if the message has been received and even read. Emails in particular can be used to send a vast amount of information and can even give links to websites or other information. Attachments to e-mails allow anything from instructions to drawings to be sent with the message.	There are few disadvantages to electronic communication, the obvious ones being no signal or a flat battery on a mobile phone and servers being down which prevent e-mails, etc. Not everyone is up to speed on the latest technology and some people are not comfortable using electronic communication. You need to make sure that the person receiving your message is able to understand how to access the information. Computer viruses can also be a problem as can security, where hackers can tap into your computer and read your e-mails and other private information. A good security set-up and antivirus software are essential.

FAQ

Why not just write the full words on a drawing?

This would take up too much space and clutter the drawing, making it difficult to read.

When working out prices or materials for quotes is it important to be exact?

No, when estimating things it is easier to round up to the nearest whole number.

Which form of communication is best?

No one way can be classed as best. Good communication depends on the circumstances – you wouldn't send a text regarding a job interview, and you wouldn't send a formal letter to a friend asking them to meet. Always choose the method of communication that works best with the situation you are in.

Check it out

1   Describe the four different methods of communication.
2   Describe the information a schedule might give you.
3   Briefly explain why drawings are used in the construction industry.
4   What do the following abbreviations stand for: DPC; hwd; fnd; DPM?
5   Sketch the graphical symbols which represent the following: brickwork; metal; sawn wood; hardcore.
6   Can you name the main types of projection that are used in building drawings?
7   What does a block plan show? Sketch an example of a block plan to show this.
8   What are dividers used for?
9   Describe the information that could be found in a drawing's title panel.
10 Explain two ways in which you can find out the height of an OBM.
11 In isometric projection, at what angle are horizontal lines drawn?
12 Describe and explain the type of information that can be found in specifications.
13 Describe the purpose of the Data Protection Act.

Getting ready for assessment

The information contained in this chapter, as well as continued practical assignments that you will carry out in your college or training centre, will help you with preparing for both your end of unit test and the diploma multiple-choice test. It will also aid you in preparing for the work that is required for the synoptic practical assignments.

Working with contract documents such as drawings, specifications and schedules is something that you will be required to do within your apprenticeship and even more so after you have qualified.

You will need to know about and be familiar with:

- interpreting building information
- determining quantities of material
- relaying information in the workplace.

To get all the information you need out of these documents you will need to build on the maths and arithmetic skills that you learnt at school. These skills will give you the understanding and knowledge you will need to complete many of the practical assignments, which will require you to carry out calculations and measurements.

You will also need to use your English and reading skills. These skills will be particularly important, as you will need to make sure that you are following all the details of any instructions you receive. This will be the same for the instructions you receive for the synoptic test, as it will for any specifications you might use in your professional life.

Communication skills have been a particular focus of this chapter, and of learning outcome 3. This chapter has explained the reasons behind recording a message and using relevant information to keep communication clear. You have also seen the benefits of positive communication over negative communication and the benefits of effective communication.

When working either professionally or on your practical assignments, you will need to communicate effectively both with the people you are working with and alongside. While studying for your qualification, you will use a range of communication methods. These will include face to face, e-mail, phone and writing. You will also need to know about the message you give to people with your body language.

The communication skills that are explained within the chapter are also vital in all tasks that you will undertake throughout your training and in life.

Good luck!

CHECK YOUR KNOWLEDGE

1   A drawing that shows you a bird's eye view of a site and its surrounding area is known as a:
    a   location drawing.
    b   component range drawing.
    c   assembly drawing.
    d   detail drawing.

2   In the title panel of a drawing there is a lot of information. Which of the following pieces of information is the most important?
    a   drawing title.
    b   architect's name.
    c   drawing number.
    d   scale.

3   A scale of 1 to 50 means that 1 mm represents:
    a   5 mm.
    b   50 mm.
    c   500 mm.
    d   5000 mm.

4   Add together the following dimensions:
    3 m + 60 cm + 9 mm.
    a   36 090 mm.
    b   3690 mm.
    c   3609 mm.
    d   30 609 mm.

5   Calculate the following: 6 m – 60 cm.
    a   54 cm.
    b   5400 mm.
    c   5.54 m.
    d   544 cm.

6   What is 250 × 8?
    a   1750.
    b   2000.
    c   2250.
    d   2500.

7   What is 12^2?
    a   144.
    b   156.
    c   160.
    d   172.

8   What is the area of a room that is 4 m long and 3.4 m wide?
    a   13 m².
    b   1.36 m².
    c   13.6 m².
    d   136 m².

9   What is an advantage of spoken communication?
    a   It is quick.
    b   It can be forgotten.
    c   There is no record of it.
    d   All of the above.

10 The written form of communication that tells us the layout of a building is known as a:
    a   specification.
    b   timesheet.
    c   drawing.
    d   schedule.

Unit 1003

Building methods and construction technology

Whatever type of building is being constructed there are certain principles that must be followed and certain elements that must be included. For example, both a block of flats and a warehouse have a roof, walls and a floor.

These basic principles are applied across all the work carried out in construction and will apply to nearly all the possible projects you could work on.

This unit contains material that supports TAP Unit 2: Set Out for Masonry Structures. It also contains material that supports the delivery of the five generic units.

This unit will cover the following learning outcomes:

- Foundations, walls and floor construction
- Construction of internal and external masonry
- Roof construction

While working through this unit, you will be practising the functional skills **FE** 1.2.1 – 1.2.3. These relate to reading and understanding information. As with the other sections in the book, if there are any words or phrases you do not understand, use a dictionary, look them up using the internet or discuss with your tutor.

Did you know?

The whole of the UK is mapped in detail and the Ordnance Survey places datum points (bench marks) at suitable locations from which all other levels can be taken

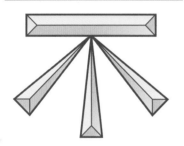

Figure 3.1 Ordnance bench mark

K1. Foundations, walls and floor construction

The majority of buildings need to be constructed so that they have a level internal surface and walls. To do this, we need to have a standard level across the whole site to ensure that all construction is being carried out to this same height. This information is given by datum points on the construction site.

Datum points

The need to apply levels is required at the beginning of the construction process and continues right up to the completion of the building.

Datum points are used to transfer levels for a range of construction jobs, including for the following:

- roads
- brick courses
- paths
- excavations
- finished floor levels.

The same basic principles are applied throughout all these jobs.

Ordnance bench mark (OBM)

OBMs are found cut into locations such as walls of churches or public buildings. The height of the OBM can be found on the relevant Ordnance Survey map or by contacting the local authority planning office. Figure 3.1 shows the normal symbol used, though it can appear as shown in Figure 3.2.

Site datum

It is necessary to have a reference point on site to which all levels can be related. This is known as the site datum. The site datum is usually positioned at a convenient height such as the finished floor level (FFL).

The site datum itself must be set in relation to some known point, preferably an OBM and must be positioned where it cannot be moved.

Figure 3.2 shows a site datum and OBM, illustrating the height relationship between them.

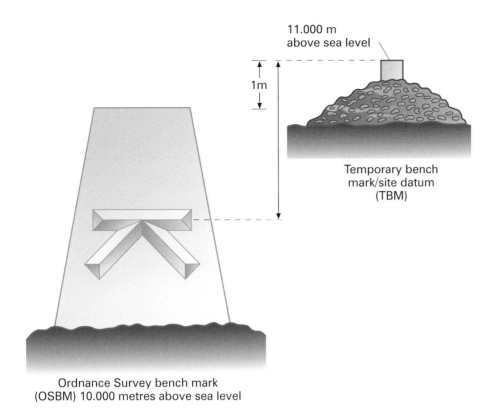

11.000 m
above sea level

1m

Temporary bench
mark/site datum
(TBM)

Ordnance Survey bench mark
(OSBM) 10.000 metres above sea level

Figure 3.2 Site datum and OBM

If no suitable position can be found a datum peg may be used. Its accurate height is transferred by surveyors from an OBM, as with the site datum. The datum peg is usually a piece of timber or steel rod positioned accurately to the required level and then set in concrete. However, it must be adequately protected and is generally surrounded by a small fence for protection (Figure 3.3).

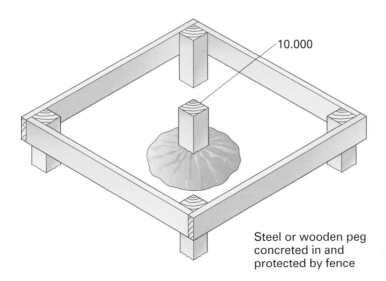

10.000

Steel or wooden peg
concreted in and
protected by fence

Figure 3.3 Datum peg suitably protected

Temporary bench mark (TBM)

When an OBM cannot be conveniently found near a site, a temporary bench mark (TBM) is usually set up at a height suitable for the site. Its accurate height is transferred by surveyors from the nearest convenient OBM.

All other site datum points can now be set up from this TBM using datum points, which are shown on the site drawings. Figure 3.4 shows datum points on drawings.

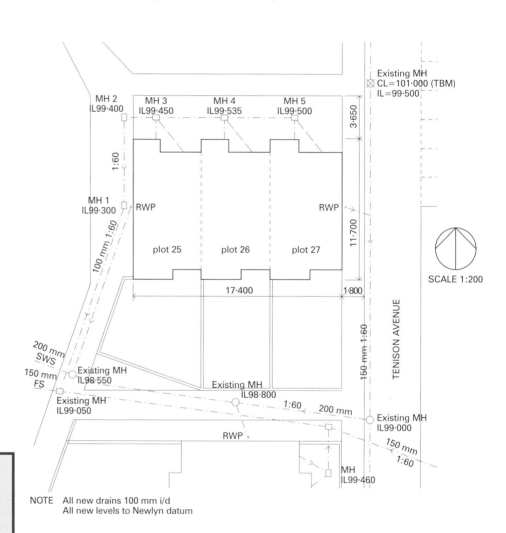

NOTE   All new drains 100 mm i/d
All new levels to Newlyn datum

Figure 3.4 Datum points shown on a drawing

Unit 1003 Building methods and construction technology

Key terms

Substructure – all of the structure below ground and up to and including the damp proof course (DPC)

Superstructure – the main building above the ground

Load bearing – something that carries a load such as a wall that supports the structure above

Substructure

All buildings will start with the **substructure**. The purpose of the substructure is to receive the loads from the **superstructure** and transfer them safely down to a suitable **load-bearing** layer of ground.

Figure 3.5 All buildings have a substructure

Did you know?

During the surveying of the soil, the density and strength of the soil are tested and laboratory tests check for harmful chemicals contained within the soil

The main material used in foundations and floors is concrete. Concrete is made up of sand, cement, stones and water.

The main part of the substructure is the foundation. When a building is at the planning stage, the entire area – including the soil – will be surveyed to check what depth, width and size of foundation will be required. This is vital: the wrong foundation could lead to the building subsiding or even collapsing.

The main type of foundation is a strip foundation. Depending on the survey reports and the type of building, one of four types of foundation will usually be used.

- **Narrow strip foundation** – the most common foundation used for most domestic dwellings and low-rise structures.

- **Wide strip foundation** – used for heavier structures or where weak soil is found.

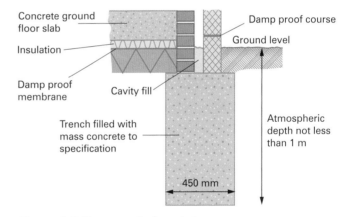

Figure 3.6 Narrow strip foundation

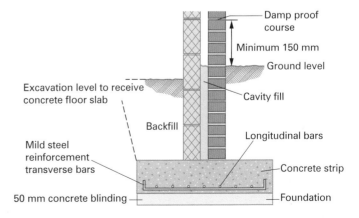

Figure 3.7 Wide strip foundation

DPC is usually made of 1000-gauge polythene, and comes in large rolls, usually black or blue in colour

Figure 3.9 Damp proof course (DPC)

Key terms

Bitumen – also known as pitch or tar, bitumen is a black sticky substance that turns into a liquid when it is heated. It is used on flat roofs to provide a waterproof seal

Slate – is a natural stone composed of clay or volcanic ash that can be machined into sheets and used to cover a roof

Figure 3.10 Ground lats

- **Raft foundation** – used where very poor soil is found. This is basically a slab of concrete that is thicker around the edges.

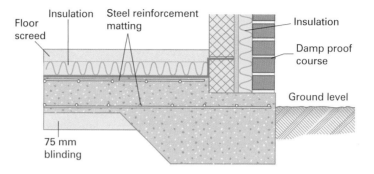

Figure 3.8 Raft foundation

Damp proof course

A damp proof course (DPC) and damp proof membranes are used to prevent damp from penetrating into a building. Flexible DPC may be made from polythene, **bitumen** or lead and is supplied in rolls of various widths for different uses.

Slate can also be used as a DPC. Older houses often have slate, but modern houses normally have polythene.

Damp proof membrane is used as a waterproof barrier over larger areas, such as under the concrete on floors.

Floors

There are two main types of floor: ground and upper.

Ground floors

There are several types of ground floor. The ones you will most often come across are:

- **Suspended timber floor** – this is a type of floor where timber joists are used to span the floor. The size of floor span determines the depth and thickness of the timbers used. The joists are either built into the inner skin of brickwork, set upon small walls (dwarf/sleeper wall), or some form of joist hanger is used. The joists should span the shortest distance; sometimes dwarf/sleeper walls are built in the middle of the span to give extra support or to go underneath load-bearing walls. The top of the floor is decked with a suitable material (usually chipboard or solid pine tongue and groove boards). As the floor is suspended, usually with crawl spaces underneath, it is vital to have air bricks fitted, allowing air to flow under the floor, preventing high moisture content and timber rot.

- **Solid concrete floor** – concrete floors are more durable and are constructed on a sub-base incorporating hardcore, damp proof membranes and insulation (Figure 3.12). The depth of the hardcore and concrete will depend on the building and will be set by the Building Regulations and the local authority. Underfloor heating can be incorporated into a solid concrete floor. You must take great care when finishing the floor to ensure it is even and level.

- **Floating floor** – these are the basic timber floor constructions that are laid on a solid concrete floor. The timbers are laid in a similar way to joists, though they are usually 50 mm thick (maximum) as there is no need for support. The timbers are laid on the floor at predetermined centres, and are not fixed to the concrete base (hence floating floor); the decking is then fixed on the timbers. Insulation or underfloor heating can be placed between the timbers to enhance the thermal and sound properties.

Upper floors

Again, solid concrete slabs can be used in larger buildings, but the most common type of upper floor is the suspended timber floor. As before, the joists are either built into the inner skin of brickwork or supported on some form of joist hanger. Spanning the shortest distance, with load-bearing walls acting as supports, it is vital that **regularised joists** are used because a level floor and ceiling are required. The tops of the joists are again decked out, with the underside being clad in plasterboard and insulation placed between the joists to help with thermal and sound properties.

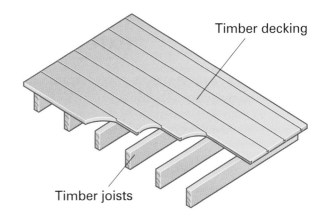

Figure 3.11 Suspended timber floor

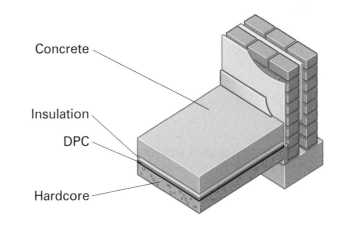

Figure 3.12 Section through a concrete floor

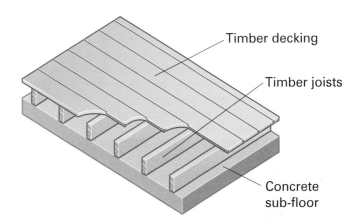

Figure 3.13 Floating floor

Key term

Regularised joists – joists that are all the same depth

K2. Construction of internal and external masonry

Bonded wall in stretcher bond

Unbonded wall

Figure 3.14 Bonded and unbonded walls

Figure 3.15 Stretcher bond wall

Bonding

Bonding is the term given to the different patterns produced when lapping the bricks to gain the most strength from the finished item. The most common type of lapping is half brick lap – better known as half bond.

If bricks were just put one on top of the other in a column, there would be no strength to the wall, and with sideways and downward pressure this type of wall would collapse. The main reasons for bonding brickwork are:

- strength
- distribution of heavy loads
- help resist sideways and downward pressure to the wall.

Other types of brick bonding are:

- stretcher bond walling
- English bond walling
- Flemish bond walling.

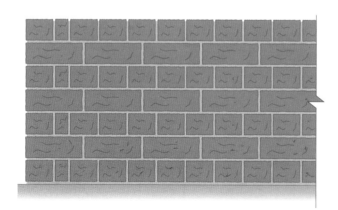

Figure 3.16 English bond wall

Figure 3.17 Flemish bond wall

Walls

External walls come in a variety of styles but the most common is cavity walling. Cavity walling is simply two brick walls built parallel to each other, with a gap between acting as the cavity. The cavity wall acts as a barrier to weather, with the outer leaf preventing rain and wind penetrating the inner leaf. The cavity is usually filled with insulation to prevent heat loss.

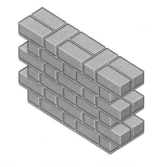

Figure 3.18 A cavity wall

Internal walls

There are several different designs of internal walls. Each has advantages and disadvantages. These methods include:

- **Blockwork** – simple blockwork, covered with plasterboard or plastered over for a smooth finish. Its disadvantage is low thermal and sound insulation.

- **Timber stud partition** – this is preferred when dividing an existing room, as it is quicker to erect. It is clad in plasterboard and plastered to smooth finish. Insulation can make the partition more fire and sound resistant. It can be load-bearing.

- **Metal stud partition** – this is similar to timber, but metal studs are used.

Figure 3.19 A solid brick wall

Wall ties

Wall ties are a very important part of a cavity wall. They tie the internal and external walls together, resulting in a stronger job. If we built cavity walls to any great height without connecting them together, the walls would be very unstable and could possibly collapse.

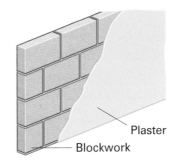

Figure 3.20 Cross-section of blockwork

Figure 3.21 Cross-section of timber stud partition

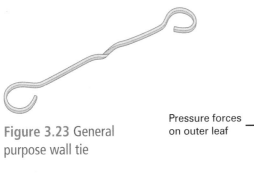

Figure 3.23 General purpose wall tie

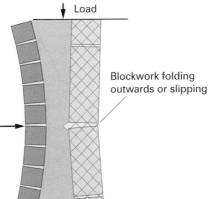

Figure 3.24 Section of walls with and without wall ties

Figure 3.22 Cross-section of metal stud partition

Figure 3.25 Stacked lintels

Unit 1003 Building methods and construction technology

Pre-cast concrete lintels

Lintels are components placed above openings in brick and block walls to bridge the opening and support the brick or blockwork above. Lintels made from concrete have a steel reinforcement placed near the bottom for strength, which is why pre-cast concrete lintels will have a 'T' or 'Top' etched into the top surface.

Mixing mortar

Mortar is used in bricklaying for bedding and jointing the bricks when building a wall. Mortar is made of sand, cement, water and plasticiser. The mortar must be 'workable' so that it can roll and spread easily. The mortar should hold on to the trowel without sticking.

Mortar can be mixed to the appropriate strength through two methods.

Gauging materials

Accurate measuring of materials to the required proportion (see Unit 1001, page 60) before mixing is important to ensure consistent colour, strength and durability of the mortar. The most accurate method of **gauging** the mortar materials is by weight. However, this method is usually only used on very large sites. The next best way to gauge the materials is by volume using a gauge box.

A gauge box is a bottomless box made to the volume of sand required (to a proportion of a bag of cement). The box is placed on a clean, level, flat surface and filled with the sand. The sand is levelled off and any spillages cleaned away. The box is then removed, leaving the amount of sand to be mixed with the bag of cement. If a gauge box is not available, a bucket could be used. A bucket is filled with sand and emptied on a clean, flat surface for the number of times specified in the proportion. A separate bucket should be used to measure the cement.

Mixing by hand

If mixing by hand, the materials should be gauged first into a pile with the cement added. The cement and sand should then be 'turned' to mix the materials together. The pile should be turned a minimum of three times to ensure the materials are mixed properly. The centre of the pile should be 'opened out' to create a centre hole.

Gradually add the water, mixing it into the sand and cement, making sure not to 'flood' the mix. Turn the mix another three times, adding water gradually to gain the required consistency.

Did you know?

Pre-cast concrete lintels come in a variety of sizes to suit the opening size

Key term

Gauging – measuring the amount of each of the components required to complete the mortar using a specific ratio such as 1:4

Did you know?

A 'shovel full' of material is not a very accurate method of gauging materials and should be avoided

Remember

It is easier to add more water than to try to remove excess water

Mixing by machine

Mixing by machine can be carried out by using either an electric mixer or a petrol or diesel mixer. Always set the mixer up on level ground.

- If using an electric mixer – the voltage should be 110 V and all cables and connections should be checked before use for splits or a loose connection (see Unit 1001, page 61). Cables should not be in contact with water and the operation should not be carried out if it is raining.

- If using a petrol or diesel mixer – make sure the fuel and oil levels are checked and topped up before starting. If using the mixer for long periods, the levels should be checked regularly so as not to run out.

Gauge the materials to be used, fill the mixer with approximately half of the water required (add plasticiser if being used). Add half the amount of cement to the water and add half of the sand. Allow to mix, and then add the remaining cement, then sand. Add more water if required, allowing at least two minutes for the mix to become workable and to ensure all the materials are thoroughly mixed together.

Once the mix has been taken out of the mixer, part fill the mixer with water and allow the water to run for a couple of minutes to remove any mortar sticking to the sides. If the mixer will not be used again that day, it should be cleaned thoroughly, either using water (and adding some broken bricks to help remove any mortar stuck to the sides) or ballast and gravel (which should then be cleaned out and the mixer washed with clean water). This will keep the mixer drum clean, and any future materials used will not stick to the drum sides so easily.

On most large sites, mortar is brought in already mixed.

K3. Roof construction

Although there are several different types of roofing, all roofs will technically be either a flat roof or a pitched roof.

Flat roofs

A flat roof is a roof with a **pitch** of 10° or less. The pitch is usually achieved through laying the joists at a pitch, or by using **firring pieces**.

The main construction method for a flat roof is similar to that for a suspended timber floor, with the edges of the joists being supported either via a hanger or built into the brickwork, or even

> **Did you know?**
>
> On cold mornings, brickwork should not be started unless the temperature is 3°C and rising

> **Remember**
>
> Never hit the drum with a hammer etc. to clean it out – this could result in costly repairs to the drum

> **Key terms**
>
> **Pitch** – the angle or slope of the roof
>
> **Firring pieces** – tapered strips of timber

Unit 1003 Building methods and construction technology

Felt – a bitumen-based waterproof membrane

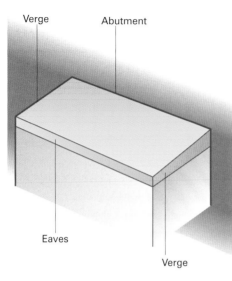

Figure 3.26 Flat roof terminology

Figure 3.27 Duo pitch roof with gable ends

a combination of both. Once the joists are laid and firring pieces are fitted (if required), insulation and a vapour barrier are put in place. The roof is then decked on top and usually plasterboarded on the underside. The decking on a flat roof must be waterproof, and can be made from a wide variety of materials, including fibreglass or bitumen-covered boarding with **felt** layered on it.

Drainage of flat roofs is vital. The edge where the fall leads to must have suitable guttering to allow rainwater to run away without draining down the face of the wall.

Pitched roofs

There are several types of pitched roof, from the basic gable roof to more complex roofs such as mansard roofs. Whichever type of roof is being fitted to a building, it will most likely be constructed in one of the following ways.

Prefabricated truss roof

As the name implies, this is a roof that has prefabricated members called trusses. Trusses are used to spread the load of the roof and to give it the required shape. Trusses are factory-made, delivered to site and lifted into place, usually by a crane. They are also easy and quick to fit: they are either nailed to a wall plate or held in place by truss clips. Once fitted, bracing is attached to keep the trusses level and secure from wind. Felt is then fixed to the trusses and tiles or slate are used to keep the roof and dwelling waterproof.

Traditional/Cut roof

This is an alternative to trusses and uses loose timbers that are cut in-situ to give the roof its shape and spread the relevant load. More time-consuming and difficult to fit than trusses, the cut roof uses rafters that are individually cut and fixed in place, with two rafters forming a sort of truss. Once the rafters are all fixed, the roof is finished with felt and tiles or slate.

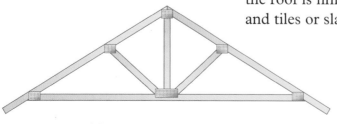

Figure 3.28 Prefabricated wooden roof truss

Figure 3.29 Individually cut rafters

Metal trusses can also be used for industrial or more complex buildings.

Finishing roofs

To finish a roof where it meets the exterior wall (eaves), you must fix a vertical timber board (fascia) and a horizontal board (soffit) to the foot of the rafters or trusses. The fascia and soffit are used to close off the roof space from insects and birds.

Ventilators are attached to the soffits to allow air into the roof space. This prevents rot. Guttering is attached to the fascia board to channel the rainwater into a drain.

> **Did you know?**
>
> As hot air rises, the majority of heat loss that occurs is through a building's roof. Insulation such as mineral wool or polystyrene must be fitted to roof spaces and ideally any intermediate floors

Roof components

Ridge

A ridge is a timber board that runs the length of the roof and acts as a type of spine. It is placed at the apex of the roof structure. The uppermost ends of the rafters are then fixed to this. This gives the roof central support and holds the rafters in place.

Figure 3.31 A ridge

Figure 3.30 Fascia and soffit in roof construction

Purlin

Purlins are horizontal beams that support the roof at the midway point. They are placed mid-way between the ridge and the wall plate. They are used when the rafters are longer than 2.5 m. The purlin is supported at each end by gables.

Firrings

A firring is an angled piece of wood, laid on top of the joints on a flat roof. They provide a fall. This supplies a pitch of around 10° or less to a roof. This pitch will allow the draining of flat roofs of water. This drainage is vital. The edge where the fall leads to must have suitable guttering to allow rainwater to run away and not down the face of the wall.

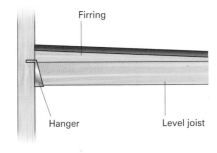

Figure 3.33 Firrings

Figure 3.32 Purlins

Did you know?

You may have heard the phrase 'batten down the hatches'. This comes from sailing when, in storms, the captain would order the crew to prevent water getting in through the hatches by sealing them with wooden strips – battening them down!

Functional skills

At the end of this unit you will have the opportunity to answer a series of questions on the material you have learnt. By answering these questions you will be practising the following functional skills:

FE 1.2.3 Read different texts and take appropriate action.

FE 1.3.1 – 1.3.5 Write clearly with a level of detail to suit the purpose.

FM 1.1.1 Identify and select mathematical procedures.

FM 1.2.1c Draw shapes.

Batten

Battens are wooden strips that are used to provide a fixing point for roof sheets or roof tiles. The spacing between the battens depends on the type of roof and they can be placed at right angles to the trusses or rafters. This makes them similar to purlins.

Some roofs use a grid pattern in both directions. This is known as a counter-batten system.

Figure 3.34 Battens on a roof

Wall plate

Wall plates are timber plates laid flat and bedded on mortar. These plates run along the wall to carry the feet of all trusses, rafters and ceiling joists. They can serve a similar role as lintels, but their main purpose is to bear and distribute the load of the roof across the wall. Because of this vital role that wall plates play in supporting the roof and spreading the pressure of its weight across the whole wall, it is important that they stay in place.

Wall plates are held in place by restraint straps along the wall. These anchor the wall plates in place to prevent movement.

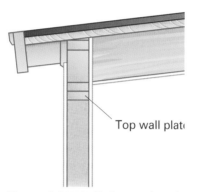

Figure 3.35 Wall plate and roof

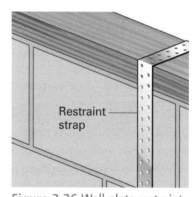

Figure 3.36 Wall plate restraint straps

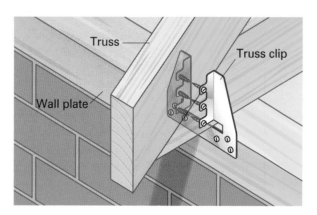

Hangers and clips

Hangers, or clips, are galvanised metal clips that are used to fix trusses or joists in place on wall plates. This anchors the trusses and joists to the wall plate, which is in turn firmly anchored to the wall. This gives the roof a stable and firm construction and helps it to avoid the pressures of wind and weather.

Figure 3.37 Truss clips

Bracing

Bracing is lengths of timber attached along the trusses. This holds the trusses in place and helps to prevent any movement in high winds. Combined with the wall plates and truss clips, bracing is a major part of ensuring the stability of the roof in all conditions.

Felt

Felt is rolled over the top of the joists to provide a waterproof barrier. It is then fastened down to provide a permanent barrier. Overlapping the felt strips when placing them will help make this barrier even more effective, as will avoiding air bubbles in the felt.

Slate and tile

Slate is flat and easy to stack. The supplier will recommend the spacing between the battens on the roof. The slate is then laid onto the battens, with the bottom of each slate tile overlapping the top of the one below. The top and bottom rows are made up of shorter slates to provide this lapping for the slate below/above. This provides waterproofing to the roof. Slate is often nailed into place.

Roofing tiles are made from concrete or clay. They are moulded or formed into a shape that allows them to overlap each other. This provides weatherproofing to the roofs, similar to the technique used for slate.

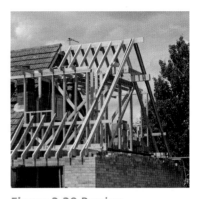

Figure 3.38 Bracing

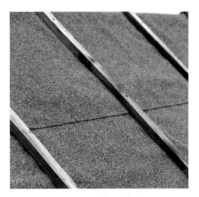

Figure 3.39 Roofing felt

Did you know?

Some roofing tiles are made from slate

Figure 3.40 Roofing slate

Figure 3.41 Roofing tiles

Flashings

Flashings are made from aluminium or lead. They are used to provide water resistance around openings in a roof such as chimneys or roof windows or when a roof butts up to an existing wall.

Metal flashing can be purchased in rolls or in sections specifically designed for certain roles – such as around a chimney.

Figure 3.42 Roof flashing

Unit 1003
Building methods and construction technology

FAQ

How do I know which type of foundation is needed?

A site survey, including taking soil samples and checking the area for tree roots, will help determine which type of foundation is used. A site survey will need to be carried out before you begin any construction work. Without it, you may overlook a potential problem with the area you are planning to build on.

How do I know which height things should be on site?

The site datum will give you a baseline to which all measurements can be taken, and the drawing will give you a measurement such as 1000 mm above site datum. You can use this information to set the relative height things will need to be on site.

Check it out

1   Explain the main purpose of substructure.

2   List the three different types of foundation. Sketch each type of these and explain the differences between them.

3   Give a brief description of external walling.

4   Explain the difference between a truss roof and a cut roof.

5   Sketch English and Flemish brick bonds.

6   Describe the purpose of a gauge box.

7   Describe the purpose of a site datum.

8   List four materials which can be used as DPC.

9   Sketch diagrams showing the cross-section of two types of internal wall.

10 Describe what mortar is used for and explain what it is made from.

Getting ready for assessment

The information contained in this chapter, as well as continued practical assignments that you will carry out in your college or training centre, will help you with preparing for both your end of unit test and the diploma multiple-choice test. It will also aid you in preparing for the work that is required for the synoptic practical assignments.

The information in this chapter will help you understand the basics of your own trade as well as the basic information on several other trade areas.

You will need to be familiar with:

- foundations, walls and floor construction
- construction in internal and external masonry
- roof construction.

It is important to understand what other trades do in relation to you and how the work they do affects you and your work. It is also good to know how the different components of a building are constructed and how these tie in with the tasks that you carry out. You must always remember that there are a number of tasks being carried out on a building site at all times, and many of these will not be connected to the work you are carrying out. It is useful to remember the communication skills you learnt in Unit 1002, as these will be important for working with other trades on site. You will also need to be familiar with specifications and contract documents, to know the type of construction work other trade workers will be doing around you on site.

For learning outcome 1 you saw how datum points work on site, and the purpose that they serve in construction. You have seen that these are vital when building a range of constructions such as roads, brick courses, paths and excavations for floor levels. You have also seen the materials used in concrete foundations and floors and the reasons for DPM and DPC.

You will need to use this knowledge to demonstrate your understanding of construction on site. Part of this is being able to sketch basic cross-sections of strip foundations and concrete floors. You will also need to be able to sketch the different types of foundation found in domestic buildings, including strip and raft concrete floor slab.

Remember, that a sound knowledge of construction methods and materials will be very useful during your training as well as in later life in your professional career.

Good luck!

CHECK YOUR KNOWLEDGE

1 What is a datum point?
 a A point from which you take all your levels.
 b A point from which you take the time and date.
 c A point that tells you which way is North.
 d A point from which you draw.

2 The most common type of foundation is:
 a raft foundation.
 b wide strip foundation.
 c narrow strip foundation.
 d none of the above.

3 The type of foundation used when the soil is very poor is called:
 a raft foundation.
 b wide strip foundation.
 c narrow strip foundation.
 d none of the above.

4 What can damp proof course be made from?
 a slate.
 b lead.
 c polythene.
 d all of the above.

5 What are wall ties used in?
 a brick walls.
 b timber stud walls.
 c block walls.
 d cavity walls.

6 What is a component placed above openings in brick and block walls called?
 a lintel.
 b cavity tray.
 c DPC.
 d wall tie.

7 What is a roof with a pitch of 10° or less called?
 a lean to roof.
 b pitched roof.
 c flat roof.
 d domed roof.

8 What is the horizontal timber board that is fixed to the foot of a rafter and finishes the roof called?
 a soffit.
 b fascia.
 c eaves.
 d gable.

9 What is the component fitted along the spine of a roof called?
 a spine board.
 b purlin.
 c ridge board.
 d soffit.

10 Name the following roofing component

Unit 1004

Produce basic woodworking joints

A carpenter or joiner must have a variety of skills at their disposal in order to be successful. Even the best work can be ruined through poor selection of materials, and the best wood in the world can be damaged if worked badly. This unit examines the basics of woodworking – how to mark out pieces for joint work, how to select the appropriate tools for the job, the basic joints you will need to know, and timber technology and selection. We will also be looking at safe working practices and tool maintenance.

This unit also supports NVQ units VR 05 Install Frames and Linings and VR 06 Install Side Hung Doors.

This unit also contains material that supports TAP Unit 5 Produce Joinery Setting Out Details. It also contains material that supports the delivery of the five generic units.

This chapter will cover the following learning outcomes:

- Marking out
- Selecting and using hand tools
- Forming basic woodworking joints
- Selecting materials

Functional skills

While working through this unit, you will be practising the functional skills **FE** 1.2.1 – 1.2.3. These relate to reading and understanding information. As with the other sections in the book, if there are any words or phrases you do not understand, use a dictionary, look them up using the internet or discuss with your tutor.

Safety tip

When painting rods or anything else, always refer to safety information on the paint tin and follow any guidelines given there

Functional skills

For setting out you will need to know about the intended use of the item. To do this you will need to refer to the item's specification. When you do this you will be practising **FE** 1.2.1 – 1.2.3, which relate to reading and understanding information including different texts and taking appropriate action, e.g. responding to advice/instructions.

K1. Know about marking out

Basic setting out

One of the most important things about setting out tools is their need for accuracy. Checking the accuracy of these tools is simple, and basic checks that can be done on the most common tools are described below:

- Measuring tools (tapes, rules) – simply measure a distance and check it with another measuring device
- Squares – mark a line then reverse the square and check that it lines up with the marked line
- Gauges – gauge a line then check the distance with a measuring tool.

At the end of this section you will understand:

- the principles of a setting out rod and its uses
- the purpose of a cutting list.

Setting out rod

A setting out rod will usually be a thin piece of plywood, hardboard or MDF, on which can be drawn the full size measurements of the item to be made. It is quite often painted white in order to aid the clarity of drawing.

Rods can be used time and time again, simply by repainting the surface upon completion of a task. If marked rods are to be kept for reuse they must be referenced and stored safely.

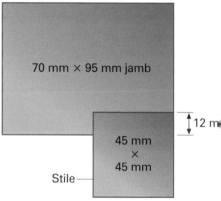

70 mm × 95 mm jamb

12 m

45 mm
×
45 mm

Stile

Stage 1

Figure 4.1 White setting out rod for sma four-pane sash

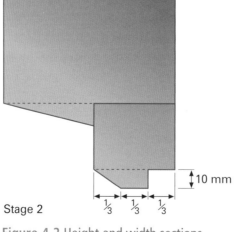

Stage 2

Figure 4.2 Height and width sections

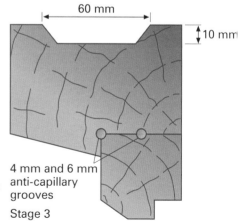

60 mm

10 mm

4 mm and 6 mm anti-capillary grooves

10 mm

Stage 3

Figure 4.3 Rod with critical dimensions for a single-panel glazed door

Upon receipt of scale drawings, specification and any on-site measurements, the **setter out** will produce a full size, horizontal and vertical section through the item by drawing it on a setting out rod. See Figure 4.1.

Elevations may also be drawn on setting out rods. This is particularly valuable for shaped or curved work, as the setter out can get a 'true' visual image of a completed joinery item.

Although rods are marked up full size, certain critical dimensions can be added as a check against any errors or damage to the rod. These are usually:

- **sight size** – the size of the innermost edges of the component (usually the height and width of any glazed components and, therefore, sometimes referred to as 'daylight size')

- **shoulder size** – the length of any member between shoulders of tenons

- **overall size** – the extreme length and width of an item.

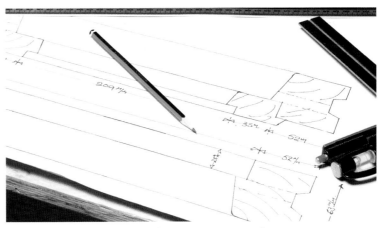

Figure 4.4 Rod marked up for a casement window

Developing drawn components

When producing workshop rods an inexperienced or apprentice joiner can sometimes have problems when building up a detailed section of timber. To overcome this, use the following step-by-step guidelines.

Figure 4.5 Step 1 Draw the components as a rectangular section

Functional skills

Using scale drawings will allow you to practise **FM** 1.2.1c, d and i, which relate to using formulae to calculate perimeters and areas. This information is important in understanding the amount of materials required to carry out a building task.

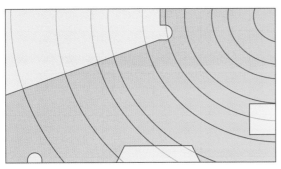

Figure 4.6 Step 2 Add any rebates, grooves and mouldings

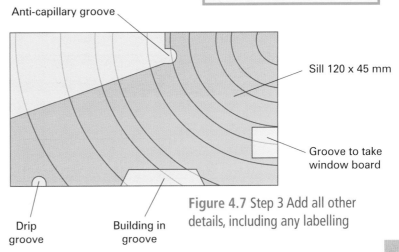

Anti-capillary groove

Sill 120 x 45 mm

Groove to take window board

Drip groove

Building in groove

Figure 4.7 Step 3 Add all other details, including any labelling

Cutting lists

Once the setting out rod has been completed the cutting list can be compiled. The cutting list is an accurate, itemised list of all the timber required to complete the job shown on the rod.

The cutting list will need to be referred to throughout the manufacturing process. It is, therefore, good practice to include the cutting list on the actual rod wherever possible.

Although there is no set layout for a cutting list, certain information should be clearly given in all lists. It should include:

- reference for the setting out rod, i.e. rod number
- date the list was compiled
- brief job description
- quantity of items required
- component description (e.g. head, sill and stile.)
- component size, both sawn and finished (3 mm per face should be allowed for machining purposes)
- general remarks.

An example cutting list is shown in Figure 4.8.

Timber cutting list						
Job description: Two panel door			Date: 8 Sept 2010			
Quantity	Description	Material	Length	Width	Thickness	Remarks
2	Stiles	S wood	1981	95	45	Mortise/groove for panel
1	Mid rail	"	760	195	45	Tenon/groove for panel
1	Btm rail	"	760	195	45	Tenon/groove for panel
1	Top rail	"	760	95	45	Tenon/groove for panel
1	Panel	Plywood	760	590	12	/
1	Panel	"	600	590	12	/

Figure 4.8 A cutting list

Tools used for measuring and marking

Measuring and marking out tools

The main tools for measuring and marking out are:

- folding rules
- retractable steel tape measures
- metal steel rules
- pencils
- marking knife
- tri-square
- sliding bevel
- mitre square
- combination square
- gauges.

Folding rules

Folding rules are used in the joiner's shop or on site. They are normally one metre long when unfolded and made of wood or plastic. They can show both metric and imperial units.

Figure 4.9 Folding rule

Retractable steel tape measures

Retractable steel tape measures, often referred to as spring tapes, are available in a variety of lengths. They are useful for setting out large areas or marking long lengths of timber and other materials. They have a hook at right angles at the start of the tape to hold over the edge of the material. On better tapes this should slide, so that it is out of the way when not measuring from an edge.

Figure 4.10 Retractable steel tape measure

Metal steel rules

Metal steel rules, often referred to as bar rules, are used for fine, accurate measurement work. They are generally 300 mm or 600 mm long and can also serve as a short straight edge for marking out. The rule can also be used on its edge for greater accuracy.

They may become discoloured over time. If so, give them a gentle rub with very fine emery paper and a light oil. If they become too rusty, replace them.

Figure 4.11 600 mm steel rule

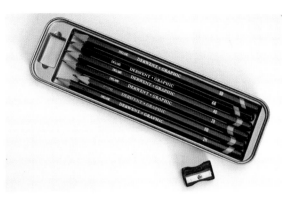

Figure 4.12 A variety of pencils

Pencils

Pencils are an important part of a tool kit. They can be used for marking out exact measurements, both across and along the grain. They must be sharpened regularly, normally to a chisel-shaped point, which can be kept sharp by rubbing on fine emery paper. A chisel edge will draw more accurately along a marking out tool, like a steel rule, than a rounded point.

Pencils are graded by the softness or hardness of the lead. B grades are soft, H grades hard, with HB as the medium grade. Increasing hardness is indicated by a number in front of the H. Harder leads give a finer line but are often more difficult to rub out. A good compromise for most carpentry work is 2H.

Marking knife

Marking knives are used for marking across the grain and can be much more accurate than a pencil. They also provide a slight indentation for saw teeth to key into.

Figure 4.13 Marking knife

Tri-square

Tri-squares are used to mark and test angles at 90° and check that surfaces are at right angles to each other.

They should be regularly checked for accuracy. To do this, place the square against any straight-edged spare timber and mark a line at right angles. Turn the square over and draw another line from the same point. If the tool is accurate the two lines should be on top of each other.

Figure 4.14 Tri-square

Sliding bevel

The sliding bevel is an adjustable tri-square, used for marking and testing angles other than 90°. When in use, the blade is set at the required angle, then locked by either a thumbscrew or set screw in the stock.

Figure 4.15 Sliding bevel

Mitre square

The blade of a mitre square is set into the stock at an angle of 45° and is used for marking out a mitre cut.

Figure 4.16 Mitre square

Combination square

A combination square does the job of a tri-square, mitre square and spirit level all in one. It is used for checking right angles, 45° angles and also that items are level.

Gauges

Gauges are instruments used to check that an item meets standard measurements. They are also used to mark critical dimensions such as length and thickness.

Figure 4.17 Combination square

Marking gauge

A marking gauge is used for marking lines parallel to the edge or end of the wood. The parts of a marking gauge include stem, stock, spur (or point) and thumbscrew. A marking gauge has only one spur or point.

Figure 4.18 Marking gauge

Mortise gauge

A mortise gauge is used for marking the double lines required when setting out mortise and tenon joints, hence the name. It has one fixed and one adjustable spur or point. Figure 4.20 shows setting of the adjustable point to match the width of a chisel.

Figure 4.19 Mortise gauge

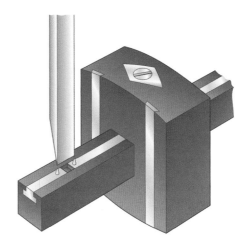

Figure 4.20 Setting mortise gauge to chisel blade width

Unit 1004 Produce basic woodworking joints

Cutting gauge

The cutting gauge is very similar to the marking gauge but has a blade in place of the spur. This is used to cut deep lines in the timber, particularly across the grain, to give a clean, precise cut (e.g. for marking the shoulders of tenons).

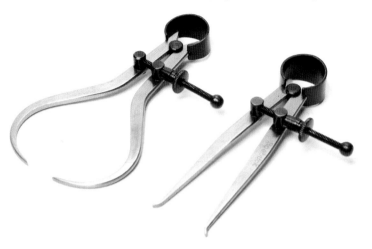

Callipers and dividers

Callipers and dividers enable accurate checking of widths and gaps. They can have a simple friction joint or knurled rod and thread. The latter are more accurate for repetitive work, as the width setting can be maintained.

Callipers are designed for either internal or external gaps. Although some come with a graduated scale it is usually better to check measurements against a steel rule.

Figure 4.21 External and internal callipers

Levelling tools

The most common levelling tool found on site is the spirit level. It has a metal body into which are fitted one or more curved glass or plastic tubes known as 'vials'. These contain a liquid and a bubble of air. They work on the principle that a bubble of gas, enclosed in a glass tube containing a liquid, will always rise to the highest possible point within the tube. There are usually marks on the glass either side of the centre at the width of the bubble. When the bubble is positioned between them the tool is level.

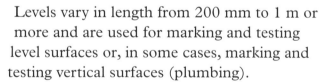

Levels vary in length from 200 mm to 1 m or more and are used for marking and testing level surfaces or, in some cases, marking and testing vertical surfaces (plumbing). Good-quality levels have adjustable tubes that can be reset should any inaccuracy develop.

Check for accuracy regularly by spanning something like a door frame and marking two level points, one at each end. Reverse the tool against the marks and see if the bubble is still level. If not, reset or replace the tool.

A range of electronic levels and plumb bobs are becoming available, and are based on lasers. They are often mounted on a stand and are self-levelling.

Figure 4.22 Spirit level

Laser levels work by shining a beam of light at the relevant object and can be used to transfer levels. Rotary laser levels are often used in the construction of suspended ceilings and floors and work in much the same way. The key difference is that in a rotary laser level, the laser spins very fast and projects a line around the room to which you can work.

K2. Know about selecting and using hand tools

As with all woodworking, selecting and using the right tools for the job is an important first step. This unit focuses on the tools needed to mark out and build woodworking joints. A more complete list of hand tools, including information on their selection and use, can be found in Unit 1005. Portable power tools will be addressed in Unit 1006.

Faulty hand tools can put you and your colleagues at risk of accident, so it is important to make sure that all of the tools you are using are in good condition. Check wooden tools for splinters, and sand down grips as needed. If any part of a tool is slightly loose, broken, or damaged in any way, it must be replaced. The best way to identify any type of faulty tool is by eye and experience. If you're not sure whether or not a tool is faulty, ask someone else's opinion. Make sure to report all faulty tools you find to your supervisor. They will make sure no one else uses them, and should put through an order for a replacement.

Percussion tools

Percussion tools can be grouped into:

- hammers
- mallets
- punches.

Hammers

Hammers are available in various types and sizes and are essential to the craftsman. The head has a protruding part, often called the bell, with a flat face for striking the work, an eye for fitting over the end of the shaft and a protruding part at the back, the pein, most often shaped as a claw, wedge or ball. Wooden shafts, made from a hardwood like ash or hickory, are still preferred for most work, as they absorb the shock. The head is secured to the shaft by driving a wedge or wedges into slots in the head of the shaft.

Figure 4.23 Laser level

Safety tip

Lasers can seriously damage eyesight, if not correctly used, so only use them if you are trained and have the appropriate PPE

Safety tip

If hammer heads become loose they must be re-secured immediately and any damaged shafts replaced

Did you know?

Loose hammer heads can be temporarily tightened on a wooden shaft by steeping in water to cause the shaft to swell, but must be repaired or replaced as soon as possible

Claw hammer

The claw hammer is used to drive nails into timber, with the claw available to withdraw bent or unwanted nails. It should be of the best quality to perform safely and efficiently. They come in various weights up to about 570 g. Increasingly they have steel shafts and integral heads, as much more leverage can be applied when using the claw without danger of loosening the head.

It is best to place a spare scrap of timber or hardboard under the claw as packing to protect the surface of the timber when using it. You can also get additional leverage and withdraw a nail straighter, thus causing less damage to the timber, by using thicker packing under the claw.

Figure 4.24 Claw hammer

Safety tip

Care must be taken when prising and removing large nails and the correct PPE must be worn

Cross pein hammer

The cross pein hammer has a wedge-shaped pein. The pein can have various patterns and the hammers come typically with head weights from 200 g to 570 g. The Warrington pattern with a tapered pein is favoured by bench joiners for lighter work.

Figure 4.25 Cross pein hammer

Mallets

Carpenter's mallets have a much larger head than hammers, and are usually made from hardwood (beech is most common) with an ash or hickory shaft. The head can be rectangular in section or round.

They are primarily for use with chisels but also for a variety of 'persuading' purposes, such as knocking components or material into place without causing damage.

There are also soft-faced mallets with either a rubber head or tightly rolled rawhide, glued and often loaded with lead. These are useful for material that wooden mallets may damage.

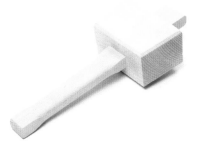

Figure 4.26 Mallet

Punches

Punches come in a wide range of shapes and sizes with different shaped tips. The type most frequently employed by a carpenter and joiner is the nail punch.

Nail punches have a square head, a knurled gripping section and a tapered point with a hollow tip. They are used to punch nail heads below the surface of the wood. The punches are available in various sizes to suit the different diameters of nails available. Smaller ones are called pin punches. The hollow tip prevents the punch slipping off the nail.

Figure 4.27 Nail punch

Push pin nail drivers can be used with pins or small nails when a hammer would be impractical. They have a built-in spring mechanism to propel the nail into the wood and can be activated one-handed.

Screwdrivers

Screwdrivers are an important part of any tool kit. There are many types and styles but they all serve a similar purpose, that is inserting and withdrawing screws. Hence, they come with different tips to match the size and type of screws available, the most common of which for wood are:

- slotted head
- Phillips cross head
- Posidrive, similar to a Phillips head but with an additional square hole in the centre for added grip by the screwdriver.

Figure 4.28 Pin punch

> **Remember**
>
> Push pin nail drivers are not for use with large diameter nails

> **Remember**
>
> Use the correct screwdriver for the job, otherwise the screw head will be damaged and either not inserted properly or difficult to remove again

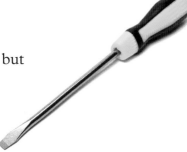

Figure 4.29 Slotted screwdriver

Figure 4.30 Posidrive screwdriver

Figure 4.31 Phillips screwdriver

> **Did you know?**
>
> The length of a screwdriver normally refers to the length of the blade only, not its overall length including handle

Screwdrivers also vary by the means used to drive screws. They can be grouped as:

- basic screwdrivers with no moving parts
- ratchet screwdrivers
- pump screwdrivers.

Basic screwdriver

Basic screwdrivers are available in various sizes from 50 mm to 400 mm. The handle is made of hardwood or unbreakable plastic. Traditional wooden handles are generally bulbous to give a good grip. Plastic ones are often moulded with flutes.

Figure 4.32 Various basic screwdrivers

Key terms

Ratchet – a one-way mechanism which uses a **pawl** to bite the sloping teeth of a wheel or bar

Pawl – a pivoted bar designed to connect to the teeth of a ratchet wheel

Screwdrivers to fit slotted screws may have tips that flare out, flared but then ground so that they are narrower towards the point, or parallel (i.e. the same width as the shank). Those made to fit other screw types are generally parallel.

Ratchet screwdriver

Ratchet screwdrivers have a ratchet mechanism built into the handle so that when turned one way the blade locks and acts like a basic screwdriver. When turned the other way the ratchet allows the handle to turn but leaves the blade in place.

There is a means to reverse the ratchet. Hence, screws can be driven in or withdrawn without changing grip on the handle. They often have interchangeable tips for different screw types, and can also be locked in one position so that they act like a basic screwdriver until the screw is sufficiently deep in the wood for the ratchet to work.

Figure 4.33 Ratchet type screwdriver

Pump action screwdrivers

Pump action screwdrivers have spiral grooves down the length of the screwdriver shank. There is a cylinder as an extension of the handle, which can move over the grooves against an internal spring.

When pressure is applied to the handle it tries to follow the grooves and, if the handle is held still, the blade is forced to turn. When pressure is released the spring in the handle allows it to return to its starting point, leaving the blade in place and ready to pump the handle down again without changing grip.

This is similar to a ratchet screwdriver and, hence, these screwdrivers are often known as spiral ratchet screwdrivers. Like them, the mechanism can also be locked in place to operate like a basic screwdriver, and normally have a chuck to take different bits.

Figure 4.34 Pump action screwdriver

The pump action allows screws to be driven and removed quickly. Therefore they are useful in repetitive operations, though it is generally advisable to drill a pilot hole unless screwing into very soft wood.

Lever tools

Lever tools are known by various names (e.g. wrecking bar, crowbar, nail bar or tommy bar) and are designed for prising items apart, levering up heavy objects or pulling large nails.

Figure 4.35 Wrecking bar

K3. Know about forming basic woodworking joints

At the end of this section you should be able to:

- understand simple jointing methods used on doors and windows
- identify the main joints used in the assembly of units and fitments
- state the correct jointing methods used on a common staircase.

The joints you will learn about in this unit are:

- housing
- mortice and tenon
- lengthening
- bridle
- halving
- dovetail
- angled
- butt and edge

Functional skills

To choose the correct joint for a job, you will need to use **FM** 1.1.1 – Identifying and selecting mathematical procedures.

Joints used on doors and windows

During the manufacture of doors and windows the mortise and tenon joint is extensively used. The type of mortise and tenon will depend on its location. Examples of this joint are described over the next few pages.

Through mortise and tenon

In a through mortise and tenon joint a single rectangular tenon is slotted into a mortise. See Figure 4.36.

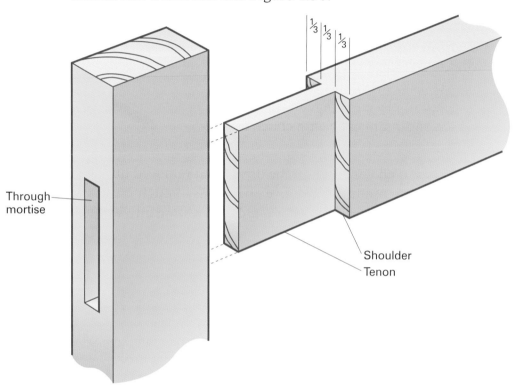

Figure 4.36 Through mortise and tenon

Stub mortise and tenon

In a stub mortise and tenon joint the tenon is stopped short to prevent it protruding through the member. See Figure 4.37.

Haunched mortise and tenon

In a haunched mortise and tenon joint the tenon is reduced in width, leaving a shortened portion of the tenon protruding which is referred to as a haunch. See Figure 4.38. The purpose of the haunch is to keep the tenon the full width of the timber at the top third of the joint. This will prevent twisting. A haunch at the end of the member will aid the wedging-up process and prevent the tenon becoming **bridled**.

Key term

Bridled – an open mortise and tenon joint. A tenon that has bridled is one that has no resistance and is therefore not secure

148

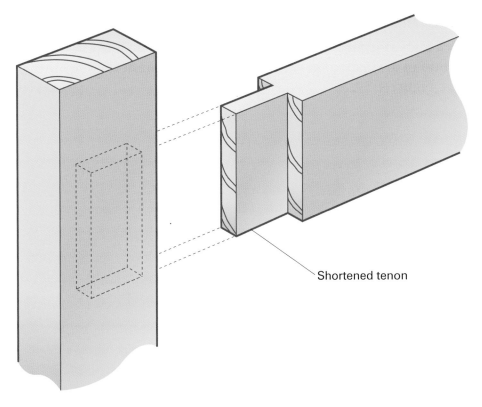

Shortened tenon

Figure 4.37 Stub mortise and tenon

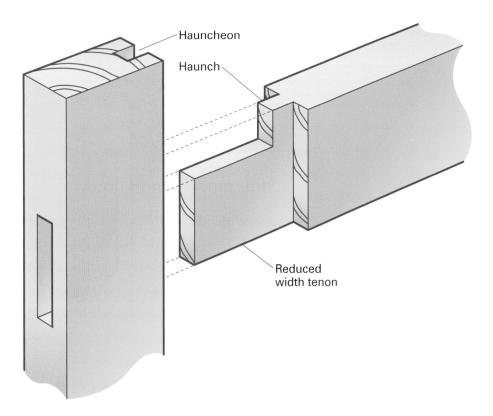

Hauncheon

Haunch

Reduced width tenon

Figure 4.38 Haunched mortise and tenon

Twin mortise and tenon

In a twin mortise and tenon joint the haunch is formed in the centre of a wide tenon, creating two tenons, one above the other. See Figure 4.39.

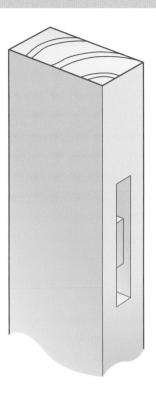
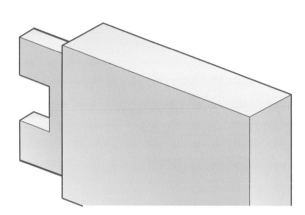

Figure 4.39 Twin mortise and tenon

Double mortise and tenon

For a double mortise and tenon, two tenons are formed within the thickness of the timber. See Figure 4.40.

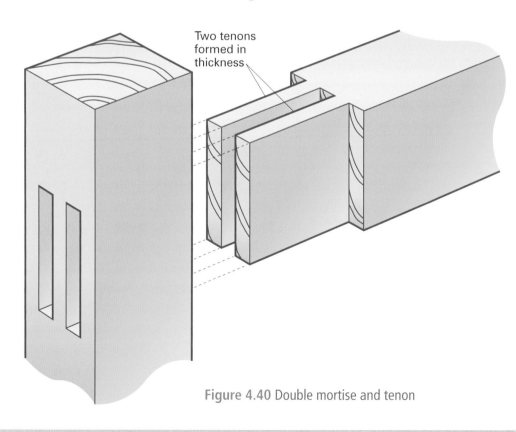

Two tenons formed in thickness

Figure 4.40 Double mortise and tenon

Stepped shoulder joint

Used on frames with rebates, a stepped shoulder joint has a shoulder stepped the depth of the rebate. This joint can also be combined with haunched, twin or stub tenons. See Figure 4.41.

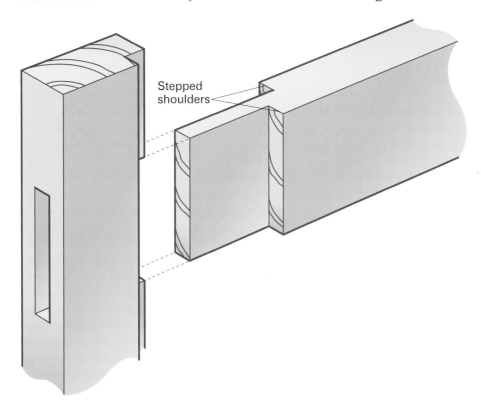

Stepped shoulders

Figure 4.41 Stepped shoulder joint

Twin tenon with twin haunch

A twin tenon with twin haunch joint is used on the deep bottom rails of doors. See Figure 4.42.

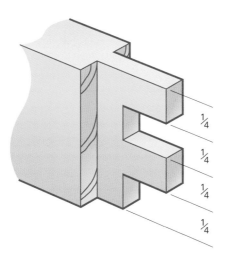

¼
¼
¼
¼

Figure 4.42 Twin tenon with twin haunch

Functional skills

As well as your basic mathematical processes, **FM** 1.2.1h refers to solving simple problems involving ratio or direct proportion. Remember this is crucial when working out the relationship between two or more values, e.g. 1:8 or 6:1:1.

Basic rules on mortise and tenon joints

The proportions of mortise and tenon joints are very important to their strength. Some basic rules follows:

- Tenon width should be no more than five times its thickness. This prevents shrinkage and movement in the joint. If it must be more than five times, a haunch should be introduced.

- The tenon should be one-third of the thickness of the timber. If a chisel is not available to cut a mortise at one-third, the tenon should be adjusted to the nearest chisel size.

- When a haunch is being used to reduce the width of a tenon, about one-third of the overall width should be removed. The depth of a haunch should be the same as its thickness.

- Although a tenon should be located in the middle third of a member, it can be moved either way slightly to stay in line with a rebate or groove.

Dovetail joints

Dovetail joints should have a slope (sometimes called the pitch) of 1:6 for softwoods, or 1:8 for hardwoods. If the slope of the dovetail is excessive the joint will be weak due to short grain. If the slope is insufficient the dovetail will have a tendency to pull apart. The slope (or pitch) is shown in Figure 4.43.

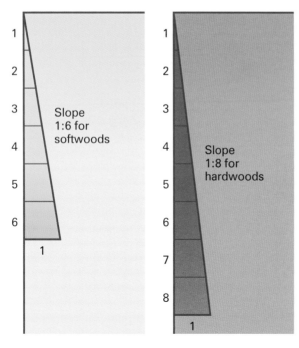

Figure 4.43 Slope of a dovetail joint

Figure 4.44 Through dovetail joint

Joints used in units and fitments

During the design and setting out of units and fitments the most common joint used is the mortise and tenon, but these are not the best if there are forces likely to try to pull the joint apart. These are called **tensile forces**.

Parts of a unit or fitment subject to such forces must incorporate a joint design that will allow for this. A drawer on a unit is often subject to tensile forces, so a dovetail joint would be used.

The two most common types of dovetail joint are through and lapped. A through dovetail joint is shown in Figure 4.44 and a lapped dovetail in Figure 4.45.

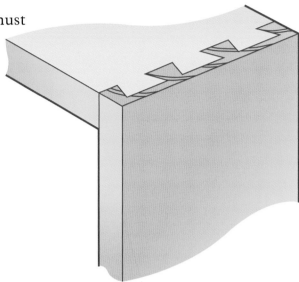

Figure 4.45 Lapped dovetail joint

Correct jointing methods on staircases

The most common joint used in staircase construction is a **stopped housing joint**. This joint is used to locate or house the tread and riser of a step into the string. It will be stopped at the nosing of the tread. The minimum housing depth is 12 mm. See Figure 4.46.

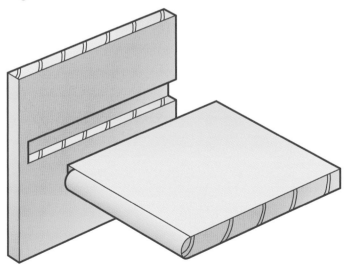

Figure 4.46 Stopped housing joint (staircase)

When the string of a stair meets a newel post, a stubbed and haunched mortise and tenon joint is used, as shown in Figure 4.47. More information on this type of joint can be found earlier in the chapter.

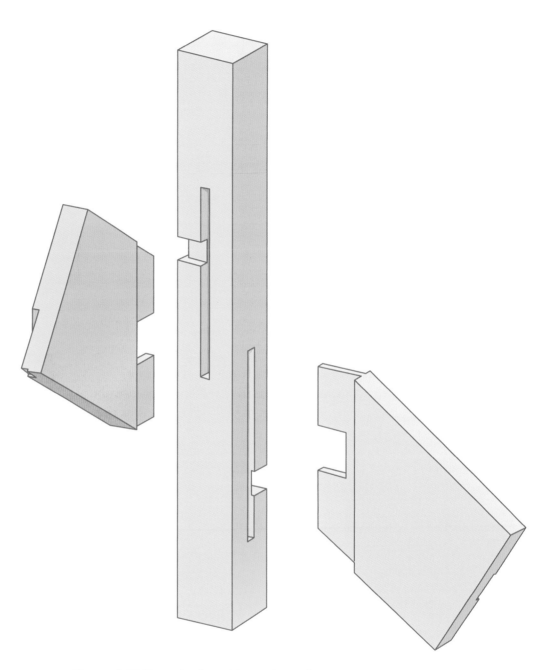

Figure 4.47 Housed string with mortise and tenon

Lengthening joints

Also known as a lap joint, a lengthening joint is similar to a halving joint. It is used to join two pieces of timber along the length. This type of joint is used in roofing for wallplates and can be used whenever the lengths of timber you have are not long enough for the task at hand.

To maintain strength the jointed section should be a minimum of 1 ½ x the width of the timber.

Bridle joints

A bridle joint is similar to a mortise and tenon, except it is reversed. Bridle joints can be used for the same purposes as mortise and tenon joints.

Halving joints

Halving joints, also known as lap joints, are used where two components meet. Half of each component is removed to allow the two pieces to form a flush joint. Halving joints are relatively simple and provide a large glue surface, giving a strong joint.

There are a number of more complex halving joints, such as the mitred halving (used at corners) or the dovetail halving.

Angled joints

Angled joints are joints in which when two pieces of timber meet and result in a change of angle. The most common angle joint is a mitre joint but they can be any angle. Pleasing to the eye but not a traditionally strong joint as there is only slightly more surface glue area than a butt joint, but the angle joint can be strengthened with mechanical fixings.

Butt and edge joints

The simplest of joints is the butt joint, which is where two pieces are simply butted together. It is a weak joint with minimal surface area for glue, but a butt joint can be strengthened with mechanical fixings.

An edge joint is simply a butt joint that runs along the edge of the components used, giving a wider component. Edge joints are usually strengthened by biscuits or dowels.

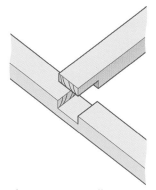

Figure 4.48 Bridle joint

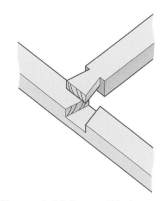

Figure 4.49 Dovetail halving joint

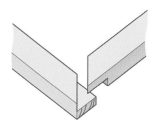

Figure 4.50 Mitre halving joint

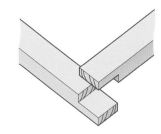

Figure 4.51 Corner halving joint

Adhesives

Adhesives, able to provide long-term fixing, are a relatively recent development within the building industry but already there are numerous products available. These range from dry lining adhesive through to expanding foam and trade name products such as No Nails™.

With any adhesive, it is important to ensure that you are using it correctly. Not all adhesives are suitable for exterior work, and not all adhesives are suitable for all materials. Contact adhesives must be allowed to nearly set before the two surfaces are introduced, or the bond will not hold.

There are also safety implications surrounding the correct use of adhesives, as they are chemically-based. Manufacturer information sheets and instruction manuals will describe the appropriate use of the adhesive as well as any information relevant to COSHH.

Figure 4.52 A selection of adhesives

Timber defects

Defects are faults that are found in timber. Some present a serious structural weakness in the timber, others do little more than spoil its appearance.

Defects can be divided into two groups:

1. seasoning defects
2. natural defects.

Seasoning defects

Seasoning defects can be further divided into:

- bowing
- springing
- winding (or twist)
- cupping
- shaking
- collapse
- case hardening.

Bowing

Bowing is usually caused by poor stacking during seasoning. It is a serious defect, causing good timber to be suitable only for use in short lengths. See Figure 4.53.

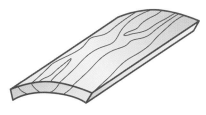

Figure 4.53 Bowing

Springing

Springing is an edgeways curvature of a board. It is usually caused by the release of internal stresses during seasoning. See Figure 4.54.

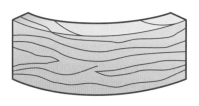

Figure 4.54 Springing

Winding (or twist)

Winding, also known as twist, is very serious as it restricts the use of the timber to short lengths. It is caused by poor seasoning and poor stacking. See Figure 4.55.

Figure 4.55 Winding (or twist)

Cupping

Cupping is very common in flat sawn boards. It occurs through shrinkage of the timber when drying. See Figure 4.56.

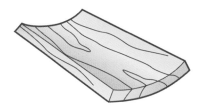

Figure 4.56 Cupping

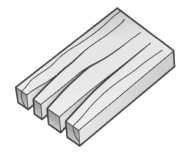

Figure 4.57 Shaking

Shaking

Shaking is caused by the board being dried too rapidly. It is particularly common at the ends of boards, spreading along the grain. See Figure 4.57.

Collapse

Collapse is a rare defect caused by too rapid drying in the early stages of seasoning. The moisture is drawn out too rapidly causing dehydrated cells to collapse. See Figure 4.58.

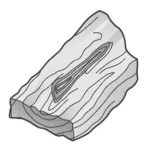

Figure 4.58 Collapse

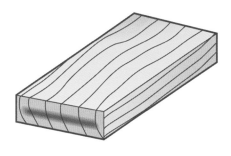

Figure 4.59 Case hardening

Case hardening

Case hardening is caused by too rapid drying, resulting in the outside cells of the timber drying and hardening, sealing off the moisture in the central part of the board. See Figure 4.59.

Unit 1004 Produce basic woodworking joints

Natural defects

Natural defects can be further divided into:

- heart shakes
- cup shakes
- star shakes
- knots.

Heart shakes

Heart shakes are usually the result of disease or over-maturity of the tree. The shakes radiate from the centre of the log and are caused by internal shrinkage. See Figure 4.60.

Cup or ring shakes

Cup shakes, also known as ring shakes, are caused by a separation of the annual rings and are usually due to a lack of nutrient or twisting of the tree in high winds. In bad cases economic conversion of the log is very difficult. See Figure 4.61.

Star shakes

Star shakes are radial cracks which occur around the outside of the log. They are caused by shrinkage at the outside of the log whilst the middle remains stable. This is usually because the log has been left too long before conversion. See Figure 4.62.

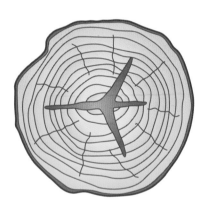

Figure 4.60 Heart shakes

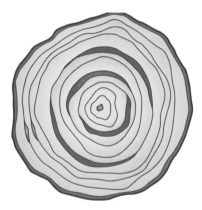

Figure 4.61 Cup shakes

Figure 4.62 Star shakes

Knots

Knots mainly occur in softwood and mark the origin of a branch in the tree. Knots are termed either 'dead' or 'live' depending on the condition of the branch which caused it. Dead knots are often loose. Small live knots aren't really a problem. Small dead knots and large knots, either dead or live, are a serious structural weakness.

Types of knot are shown in Figures 4.63 to 4.66.

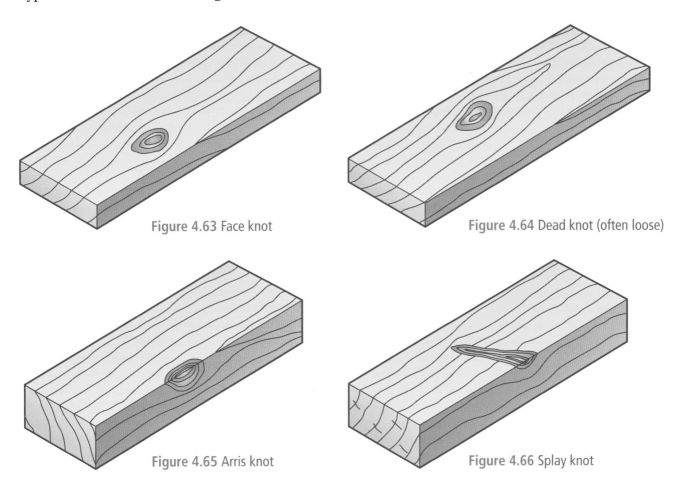

Figure 4.63 Face knot

Figure 4.64 Dead knot (often loose)

Figure 4.65 Arris knot

Figure 4.66 Splay knot

K4. Know about selecting materials

Knowing the right materials to select for the task at hand is one of the most important skills in the arsenal of a good carpenter or joiner. Different types of timber have different properties, making them suitable for completely different tasks. Balsa wood, which is popular in modelling, is completely unsuitable for structural carcassing. Equally, it would be foolish to use oak when constructing a lightweight model airplane.

Timber conversion

Conversion is the sawing of a log into boards or planks ready for use by the carpenter or joiner. How the timber is converted directly affects its usefulness.

Softwoods are nearly always sawn in their country of origin, while hardwoods are often imported in log form and converted by the timber merchant, sometimes to customer requirements.

> **Did you know?**
>
> Timber contains millions of tiny cells which normally contain a mixture of air, water and other chemicals

Methods of conversion

There are four main methods of conversion:

- through and through
- tangential
- quarter
- boxed heart.

Through and through sawn timber

Also known as flat slab or slash sawing, through and through sawing is the simplest and most economical method of converting timber. See Figure 4.67. Although there is very little wastage with this method, the majority of the boards produced are prone to a large amount of shrinkage and distortion.

Tangential sawn timber

The tangential sawn timber method of conversion is used to provide floor joists and beams, as it gives the strongest timber. See Figure 4.68. It is also used on pitch pine and douglas fir for decorative purposes to produce 'flame figuring' or 'fiery grain'.

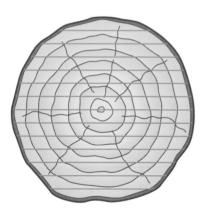

Figure 4.67 Through and through sawn timber

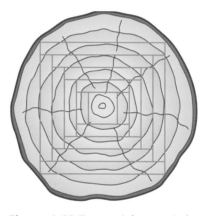

Figure 4.68 Tangential sawn timber

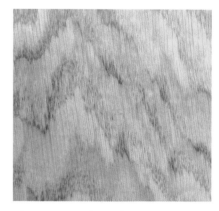

Figure 4.69 Flame figuring

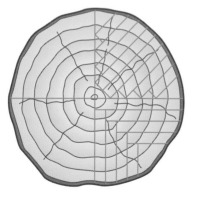

Figure 4.70 Quarter sawn timber

Quarter sawn timber

Quarter sawn timber, shown in Figure 4.70, produces the best-quality timber. However, it is also the most expensive, both in time involved and material wastage. It produces the greatest quantity of 'rift' or 'radial sawn' boards, which are generally superior for joinery purposes.

Boxed heart sawn timber

Boxed heart sawn timber is a type of radial sawing that is used when the heart of the tree is rotten or badly shaken. It is sometimes known as floor boarding sawing, as the boards are ideal for this purpose because they are hard-wearing and do not distort. See Figure 4.71.

Seasoning of timber

Timber from newly felled trees contains a high proportion of moisture in the form of sap, which is made up of water and minerals drawn from the soil. Most of this water has to be removed by some form of drying, which is called seasoning.

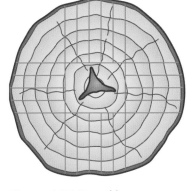

Figure 4.71 Boxed heart sawn timber

The main reasons for seasoning are:

- to make sure that shrinkage occurs before the timber is used
- to make sure that the moisture content of the timber is below the 'dry rot' safety line of 20 per cent (discussed later in this chapter)
- to make sure that dry timber is used
- dry timber is stronger
- seasoned timber is less likely to split or distort
- wet timber will not accept glue, paint or polish.

The timber should be dried to a moisture content that is similar to the surrounding atmosphere in which it will be used. See Table 4.1.

Table 4.1 Moisture content table

Timber location	Moisture content (per cent)
Carcassing timber (joists, etc.)	18–20
External joinery	16–18
Internal timber where there is a partial intermittent heating system	14–16
Internal timber where there is a continuous heating system	10–12
Internal timber placed directly over, or near, sources of heat	7–10

There are two main methods of seasoning timber:

- natural, normally called 'air seasoning'
- artificial, normally called 'kiln seasoning'.

Air seasoning

For air drying the timber is stacked in a pile in open-sided, covered sheds, which protect the timber from rain but still allow a free circulation of air. A moisture content of 18–20 per cent can be achieved in a period of 2–12 months.

Kiln seasoning

Most timber that is used is kiln seasoned. If done correctly the moisture content of the timber can be reduced without causing any timber defects.

Depending on the size of the timber, the length of time the timber needs to stay in the kiln varies between two days and six weeks.

There are two types of kiln in general use:

- compartment kiln
- progressive kiln.

A compartment kiln is shown in Figure 4.72. It is usually a brick or concrete structure, in which the timber remains stationary during the drying process.

The drying of the timber depends on three factors:

- air circulation supplied by fans
- heat, usually supplied by heating coils
- humidity, which is raised by steam sprays.

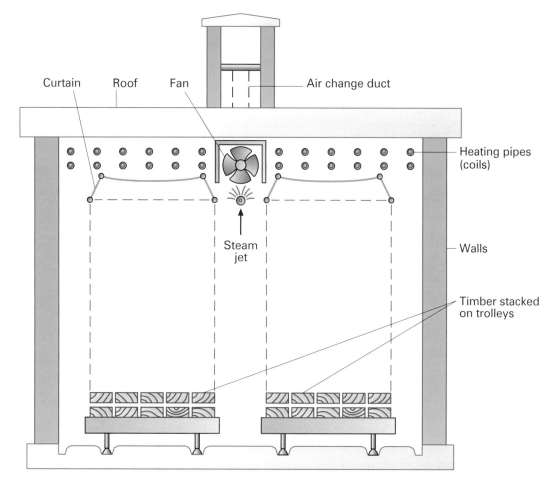

Figure 4.72 Compartment kiln

In a progressive kiln the timber is stacked on trolleys, which pass slowly through a long chamber, with gentle changes to heat and humidity as it moves from one end of the chamber to the other.

There are three clear advantages of using kiln, as opposed to air drying:

1. the speed at which the seasoning can be completed
2. the facility to dry timber to any desired moisture content
3. the sterilising effect of the heated air upon fungi and insects in the timber, lessening the likelihood of fungal or insect attack.

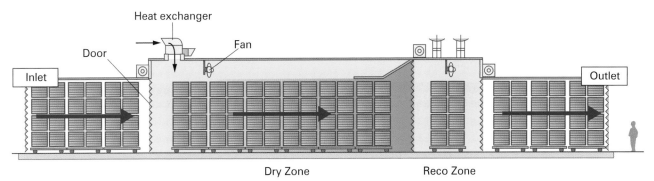

Figure 4.73 Progressive kiln

Materials storage

The storing of materials was covered in Unit 1001 on pages 37–59. The most common types of wood and sheet materials are:

- softwood (red deal and white wood)
- hardwood (oak, ash, mahogany)
- timber manufactured boards (chipboard, plywood, medium density fibreboard (MDF) and blackboard).

Timber technology

Classification of timber

Timber is classified as either hardwood or softwood. This can sometimes be confusing, as not all hardwoods are physically hard or softwoods soft. For example, the balsa tree is classified as hardwood although it is very soft and light. The wood of a yew tree, classified as softwood timber, is harder than most hardwoods.

Key terms

Botanical – the classification of trees based on scientific study

Deciduous – the name given to a type of tree that sheds its leaves every year

Evergreen – a type of tree that keeps its leaves all year round

Hardwood and softwood refer to the **botanical** differences and not to the strength of the timber. Generally speaking, hardwood trees are **deciduous**, broad-leafed, with an encased seed. Softwood trees are usually **evergreen** with needles and seeds held in cones.

Identification of timber

One of the best ways of becoming familiar with, and being able to identify, a reasonably wide range of timber species, is to form a personal collection. Small, matchbox-size samples are ideal for this.

The key to learning and memorising timber species is colour. Once timbers have been grouped together, identification becomes a process of elimination. Tables 4.2 (softwoods) and 4.3 (hardwoods) outline the general appearance, distribution and uses of timber species most commonly used.

Table 4.2 Commonly used softwoods

Name	Main Sources	Identification	Properties/Description	Uses
Douglas fir	USA, Canada		Straight grained and resilient: easy to work with by hand or machine. One of the hardest softwoods can take heavy, continuous wear. A high resistance to acids and decay. Has good gluing and high insulation qualities	First class joinery, light and heavy structural work, glued laminated work and plywood
Larch	Europe		Very strong and durable, average to work with and sawing and machine can leave a decent finish but loose knots can be troublesome	Fencing, gate posts, garden furniture, railway sleepers
Pitch pine	South USA		Similar strength to Douglas fir. Works moderately easily, but the resin is often troublesome, tending to clog saw-teeth and cutters, and to adhere to machine tables. A good finish is obtainable, and the wood can be glued satisfactorily, takes nails and screws well	Shipbuilding, polished softwood joinery and church furniture

Table 4.2 Commonly used softwoods (cont.)

Name	Main Sources	Identification	Properties/Description	Uses
Redwood (commonly known as Pine)	Europe		Moderately strong for its weight with average durability. The timber works easily and well with both hand and machine tools, but ease of working and quality of finish is dependent upon the size, number of knots, and amount of resin. The wood is capable of a smooth, clean finish, and can be glued, stained, varnished and painted. Takes nails and screws well	Depending on quality it can be used for interior or exterior work and can be used for most tasks such as carcassing and finish joinery
Whitewood (also known as European Spruce)	Europe		Similar to redwood or scots pine in strength and durability. Good to work with by both hand and machine, takes glue, nails and screws well and can produce a good finish	Similar uses to redwood
Western red cedar	Canada, USA		Not as strong as redwood but has naturally occurring oils which prevent insect attack. Non-resinous and straight grained. Good workability with both hand and machine. Doesn't need treating as the wood will stand up to severe weather and will turn a silvery colour when exposed	Externally for good quality timber buildings, saunas, etc.

Table 4.3 Commonly used hardwoods

Name	Main Sources	Identification	Properties/Description	Uses
Ash	Europe		Straight grained, very tough and flexible. Although tough, ash works with machines quite well, and has a reasonably smooth finish. It can be glued, stained, and polished and takes nails and screws well	Furniture, boat building, sports equipment, tool handles, etc.
Oak	Europe		Very strong with English oak being the strongest. Good resistance to bending and shearing and workability is quite tough but can be planed to a good finish. Susceptible to fungal attack and ironwork should not be placed onto untreated oak as it will disfigure the timber and leave a bluish stain. Glues well but difficult to screw and nail	High-class joinery, panelling, doors exposed roofing, etc.
Beech	Europe		Hard, close grained and durable with a fine texture. Works fairly well by both hand and machine, and is capable of a good smooth surface. Takes glue, stains and polish well and can produce an excellent veneer	Furniture, kitchen utensils, wood block floors, etc.

Table 4.3 Commonly used hardwoods (cont.)

Name	Main Sources	Identification	Properties/ Description	Uses
Mahogany	Africa		Interlocked grain, reasonably strong for its weight and moderately resistant to decay. Fairly easy to work with hand and machine, takes glues, finish, nails and screws well	High-class joinery, furniture, boat building and plywood veneers
Mahogany	Cuba		Extremely strong for its weight, it also has an interlocking grain. Very good to work with and takes glue, nails, etc. very well and can be polished to an excellent finish	As for African Mahogany but considered to be superior
Maple rock	Canada, North East USA		Fine grained with excellent finishing qualities and average durability. Can be difficult to work with and is hard to nail or screw	Panelling, flooring, furniture and snooker cues
Sapele	West Africa		Harder than mahogany with similar strength properties to oak. Works fairly well with hand and machine tools, but the interlocked grain is often troublesome in planing and moulding. It takes screws and nails well, glues satisfactorily, stains readily, and takes an excellent polish	Furniture, veneers, etc.

Table 4.3 Commonly used hardwoods (cont.)

Name	Main Sources	Identification	Properties/Description	Uses
Teak	India, Java, Thailand		Greasy to touch and resistant to both insect and fire, the timber is strong, very durable, moderately elastic, and hard, but is rather brittle along the grain. Workability can be difficult and tools soon become blunt but the wood is capable of a good finish if cutting edges are kept sharpened. It can be glued, stained and polished. It holds screws and nails reasonably well, but the wood is inclined to be brittle	High-class joinery, furniture, boat building, etc.
Walnut	Europe		A relatively tough wood with good resistance to splitting. Easy to work with hand and machine and takes glue, nail and screw well. Can be polished to an excellent finish. Staining likely if in contact with iron in damp conditions	Furniture veneers, etc

Manufactured boards

The most common types of wood-based manufactured boards are:

- plywood
- laminated board
- chipboard (particle board)
- fibre board.

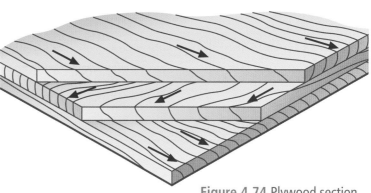

Figure 4.74 Plywood section showing grain direction

Plywood

Plywood is made from thin layers of timber called **veneers**. The veneers are glued together to form boards. There is normally an odd number of veneers with the grain alternating across and along the sheet, which gives both strength and stability to the board (see Figure 4.74).

Plywood is graded according to the type of glue used in its manufacture and also according to the situation in which it will be used. The grade is usually stamped on the board by the manufacturer. A list of grades is shown in Table 4.4.

Table 4.4 Plywood grade table

Stamp	Grade	Use
INT	Interior	Internal use only – has low resistance to humidity or dampness
MR	Moisture resistant	Has a fair resistance to humidity and dampness
BR	Boil resistant	Has a fairly high resistance in exposed conditions
WBP	Weather and boil proof	Can be used in extreme conditions under continuous exposure (boats, buildings, etc.)

Laminated boards

Laminated boards are made from strips of wood laminated together and sandwiched between two veneers. Three common varieties are shown in Figure 4.75.

At the end of this unit you will have the opportunity to answer a series of questions on the material you have learnt. By answering these questions you will be practising the following functional skills:

FE 1.2.3 Read different texts and take appropriate action.

FE 1.3.1 – 1.3.5 Write clearly with a level of detail to suit the purpose.

FM 1.1.1 Identify and select mathematical procedures.

FM 1.2.1c Draw shapes.

Blockboard
Strips are up to 25 mm wide. Good quality hardwood veneers are sometimes used.

Laminboard
Strips are 7–8 mm wide. This produces a better quality board.

Battenboard
Strips are up to 75 mm wide, producing a poorer quality board.

Figure 4.75 Laminated boards

Chipboard

Chipboard is made from compressed wood chips and wood flakes bonded with a synthetic resin glue. See Figure 4.76. There are various grades, which include those used to make floor panels.

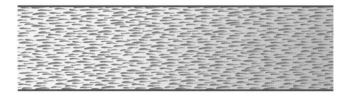

Figure 4.76 Chipboard

Fibreboard

Fibreboard is made from pulped wood that is mixed with an adhesive and pressed into sheets. It usually comes in three forms:

- hardboard
- insulation board (or softboard)
- medium density fibreboard (**MDF**).

Safety tip

Dust extraction equipment should be used or a respirator should be worn when cutting or sanding any timber, but particularly some tropical hardwoods and MDF, which can be harmful

Hardboard

Hardboard is manufactured from sugar cane pulp and available in sheets 3–6 mm in thickness. Oil-tempered hardboard offers a reasonable resistance to moisture. See Figure 4.77.

Hardboard is also manufactured with various finishes, which include:

- plastic faced
- perforated
- reeded
- enamelled.

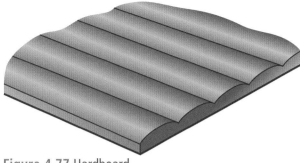

Figure 4.77 Hardboard

Insulation board (or softboard)

Also called softboard, insulation board is made from the same material as hardboard but not compressed. It is used as a wall and ceiling covering to give very good insulation. See Figure 4.78.

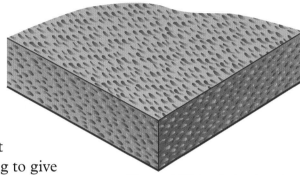

Figure 4.78 Insulation board

Medium density fibreboard (MDF)

Medium density fibreboard (MDF) is used to form skirting boards, and mouldings such as architraves. It is easy to work with both hand and powered tools. It is mainly used internally but moisture-resistant sheets are available.

Did you know?

Architrave is a name for mouldings around windows and doors

Working Life

Wayne goes to the store to get some timber for a job. When Wayne lifts the timber from the shelf, he notices that the wood is in wind (i.e. twisted).

- How could this have happened? What could have been done to prevent it from happening? Think about what causes twisting in timber and the measures you could put in place to stop it from happening. Think about the storage of the materials.

- Do you think Wayne can still use the timber? Give reasons for your answer. If he does use it what might the impacts be? What is different about the wood that might affect its use?

Functional skills

When answering questions you will need to make sure that you fully understand what they are asking. In doing this you are practising **FE** 1.2.1 – 1.2.3 which relate to reading and understanding information including different texts and taking appropriate action, e.g. responding to advice/instructions. Answering questions allows you to practise **FE** 1.3.1 – 1.3.5. Write clearly with a level of detail to suit the purpose.

FAQ

How do I choose between hardwood and softwood for a job?

The type of timber you should use for a job is usually detailed in the specifications. There are several reasons why one type of wood is chosen over another. Hardwood is usually more expensive and longer lasting than softwood. It is often grown in the hot climates of equatorial countries (e.g. African and South American countries) and is used for jobs where the wood will be visible (i.e. high-class joinery). Softwood is usually grown in countries with cooler climates. It is often cheaper than hardwood and used for jobs where the wood will be concealed (e.g. floorboards and rafters).

Why use a double tenon joint? Would it not be easier to just put in a thicker single tenon?

Yes, it would be easier but it would not be as strong. Putting in a double tenon will increase the surface area of the joint. This will give the joint a larger area for adhesion (gluing), producing a stronger, well-proportioned joint.

Where would a twin tenon with twin haunch be used as opposed to a twin tenon?

A twin tenon is ideal for the top of stairs where the newel post is thicker than the string, but a twin tenon with twin haunch should be used at the bottom of the stair, where the newel post is cut flush and the twin haunch will give additional strength.

Why would you need sawn sizes on a cutting list?

By putting sawn sizes on the cutting list the machinist will quickly be able to determine the most cost-effective sections of stock to use from the timber rack.

Check it out

1 State what a haunch is and describe why it is used.
2 Describe when a tenon with a stepped shoulder should be used and why.
3 A door stile has a finished section of 95 mm × 44 mm. Which size chisel would be used to cut the mortise?
4 What pitch should be used for a dovetail in softwood?
5 State the joints used in the following:
   a) stair tread to string
   b) string to newel post
   c) middle rail of a two panel door
   d) drawer on a kitchen unit.
6 Describe the purpose of setting out and cutting lists.
7 Explain each of the following: sight size; shoulder size; overall size.
8 What is the purpose of face and edge marks?
9 Explain how to mark out.

Getting ready for assessment

The information contained in this chapter, as well as continued practical assignments that you will carry out in your college or training centre, will help you with preparing for both your end of unit test and the diploma multiple-choice test. It will also aid you in preparing for the work that is required for the synoptic practical assignments.

The information contained within this chapter will aid you in learning how to identify and calculate the materials and equipment required for the marking and setting out of basic woodworking joints.

You will need to be familiar with:

- marking out joints
- selecting and using the correct hand tools
- selecting and using the correct materials
- forming basic woodworking joints

This unit has explained to you the reasons behind the decisions you will need to take on any practical assessment. In your synoptic test you will need to use all this knowledge you have collected in a practical way.

For example, for learning outcome three you need to know about forming basic woodworking joints. This unit has looked at common woodworking joints and described when and why they are used. This information will be vital when you read the specifications of a practical assignment, as this will instruct you on the conditions you need your joint to work to. You will use similar skills when you are working professionally with specifications.

This unit has also identified the key hand tools you will need to work with when carrying out this work. You have also seen the properties and uses of adhesives and some of the flaws and defects that can occur in types of timber. All this information will be vital when you are working with tools and materials. It will also help you to select the correct tools and materials for each job, and to make sure that you are using them correctly at all times.

Before you start work on the synoptic practical test it is important that you have had sufficient practice and that you feel that you are capable of passing.

On approaching the test it is best to have a plan of action and a work method that will help you. You will also need a copy of the required standards, any associated drawings and sufficient tools and materials. It is also wise to check your work at regular intervals. This will help you to be sure that you are working correctly, and help you to avoid problems developing as you work.

Your speed at carrying out these tasks will also help you to prepare for the time limit that the synoptic practical task has. But remember, do not try to rush the job as speed will come with practice and it is important that you get the quality of workmanship right.

Always make sure that you are working safely throughout the test. Make sure you are working to all the safety requirements given throughout the test and wear all appropriate personal protective equipment. When using tools, make sure you are using them correctly and safely.

Good luck!

CHECK YOUR KNOWLEDGE

1 The type of timber used for stud partitions and framework is:
   a carcassing.
   b joinery grade.
   c hardwood.
   d sheet material (plywood).

2 Plywood and other sheet materials should be stored:
   a against a wall.
   b flat on the floor.
   c flat on bearers.
   d in a warm room.

3 The timber shown is

   a pitch pine.
   b larch.
   c beech.
   d redwood.

4 The timber shown is

   a oak.
   b African mahogany.
   c maple rock.
   d walnut.

5 Which man-made board is made from pulped wood mixed with adhesive?
   a chipboard.
   b fibreboard.
   c laminated board.
   d plywood.

6 When do seasoning defects occur?
   a when the tree is growing.
   b when the tree is being felled.
   c when the tree is converted.
   d when the converted timber is dried out.

7 Disease or over-maturity of the tree can result in which defect?
   a heart shake.
   b cup shake.
   c ring shake.
   d loose knots.

8 Why is kiln seasoning preferred to air seasoning?
   a The speed at which the seasoning can be completed.
   b The facility to dry timber to any desired moisture content.
   c The sterilising effect of the heated air upon fungi and insects in the timber.
   d All of the above.

9 The most important thing about setting out tools is:
   a that they are kept clean.
   b their accuracy.
   c their price.
   d none of the above.

10 What is the most common joint used in staircase construction?
   a dovetail joint.
   b stopped housing joint.
   c mortise and tenon joint.
   d halving joint.

Unit 1005

Maintain and use carpentry and joinery hand tools

Without properly maintained tools, the job of a carpenter or joiner is much more difficult. As well as knowing how to maintain and store tools to get the most out of them, a carpenter or joiner must know the best – and safest – techniques for using all of the tools in their toolbox. Poorly maintained tools can damage high-quality timber, costing you both time and money in the long run. Knowing which tool to use for which job, and how to use them safely, will help reduce the risk of accident for you and your colleagues, as well as making your work more professional and efficient.

This unit also contains material that supports TAP Unit 2 Install First Fixing Components. It also contains material that supports the delivery of the five generic units.

This unit will cover the following learning outcomes:

Maintaining and storing hand tools

Using hand saws

Using hand-held planes

Using hand-held drills

Using woodworking chisels

Functional skills

When reading and understanding the text in this unit, you are practising several functional skills.

FE 1.2.1 - Identifying how the main points and ideas are organised in different texts.

FE 1.2.2 – Understanding different texts in detail.

FE 1.2.3 – Read different texts and take appropriate action, e.g. respond to advice/instructions.

If there are any words or phrases you do not understand, use a dictionary, look them up using the internet or discuss with your tutor.

Key terms

Maintenance – the care or upkeep of an object. Good maintenance keeps your tools in good condition, saving you money in the long run

Functional skills

While working through this unit, you will be practising the functional skills **FE** 1.2.1 – 1.2.3. These relate to reading and understanding information. As with the other sections in the book, if there are any words or phrases you do not understand, use a dictionary, look them up using the internet or discuss with your tutor.

K1. Know how to maintain and store hand tools

Hand tools

Although portable power tools are increasingly being used, and can complete many tasks much faster than hand tools, a good carpenter and joiner will still need to use a wide range of hand tools.

It is essential to have a basic set of hand tools. It is also good practice to extend your tool kit by purchasing good tools, as and when needed for a particular job, so building up a set of high-quality tools over time.

In this section we will look at each type of hand tool in turn, including where and how to use them. We will also address general safety measures and recommendations. Finally we will look at the protection and **maintenance** of tools, so that they give long and efficient service.

Safety first

The following are general safety rules when using any hand tools.

- Wear the correct **PPE** for the task, particularly good-quality safety glasses if there is any danger of fast-flying objects.
- Do not wear loose clothing or jewellery that could catch on tools.
- Secure all work down before working on it.
- Keep hands well away from the sharp edges of tools, especially saw teeth, chisels and drill bits.
- Never use a tool to do a job for which it was not designed.
- Make sure that tools are kept sharp and stored properly when not in use.
- Never force the tool – the tool should do the work.
- Never place any part of your body in front of the cutting edge.
- Keep work areas clean.

Summary of hand tool types

Hand tools can be grouped in the following categories, each of which is covered separately below, with illustrations:

- sawing tools
- cutting tools
- planing tools
- shaping tools
- drilling or boring tools
- holding and clamping tools.

Maintenance of hand tools

Good-quality hand tools will last a lifetime if well looked after. All metal tools need to be kept free of rust, which can be achieved by rubbing them over occasionally with an oily cloth to preserve a light film of oil.

Sharpening planes and chisels

Sharpening is best done by hand using an **oilstone**. These come in coarse, medium or fine grades, often as a combination with two of them on opposite sides. Tool sharpening should always be completed on a fine grade, the others only used for reshaping. Stones can be cleaned with a brush and paraffin and should be re-levelled occasionally by rubbing on a sheet of glass with carborundum paste. The stone should always be kept oiled during sharpening.

Honing angle

Grinding angle

Figure 5.1 Honing and grinding angles

The following procedures will enable you to sharpen the blades of planes and chisels.

Figure 5.2 Combination oilstone

Step 1 Position the blade. Holding the blade in both hands, position its grinding angle flat on the stone. This means the blade will be at an angle of 25°. Then raise the back end up a further 5°, so that the tool is now at the correct sharpened angle of 30°.

Step 2 Grind an initial burr. Slowly move the blade forwards and backwards until a small burr has formed at the cutting edge. Make sure that you use the entire oilstone surface, to avoid hollowing the stone.

Figure 5.3

Figure 5.4

Step 3 Form a wire edge. Remove the burr by holding the blade perfectly flat and drawing the blade towards you once or twice to form a wire edge on the end of the blade.

Step 4 Remove the wire edge. The wire edge can be removed by drawing the blade over a piece of waste timber. Now inspect the sharpened edge and, if you can still see a dull white line, repeat Steps 2 to 4.

Figure 5.5

Figure 5.6

Two additional tips specifically for sharpening planes are:

- Hold the blade at a slight angle to the line of the stone. This helps to ensure that the whole of the cutting edge makes contact with the stone.
- To ensure that the cap iron fits neatly with the blade, also use the oilstone to flatten and straighten its edge.

Over the years there has been a range of legislation covering the use of grinding wheels. These have all been superseded by PUWER, which is discussed on page 9 of Unit 1001.

Saw setting and sharpening

We will cover the types of saws on page 189–92. Most modern saws are of the hard-point type and cannot be re-sharpened. For these, replacement is the only option. Older saws do require periodic maintenance, the frequency depending on the amount of work they do.

There are four stages in returning a saw to tip-top condition:

- topping
- shaping
- setting
- sharpening.

For all four operations a pair of saw stocks is required to hold the saw blade tight with the teeth uppermost. These can be made by clamping two suitable lengths of softwood either side of the saw blade in a vice.

There are various saw files available, designed for the different tasks.

Topping or levelling the teeth is completed by running a flat saw file along the top of the teeth. A simple jig to help with this can be made of hardwood, with a slot cut in it to hold the file at right angles. The wood can then rub against the blade to ensure that the file remains flat on the teeth.

Shaping is undertaken with a saw file designed for the task. This restores the teeth to their original shape and size. The aim is to have all teeth with the same angle, or pitch.

Teeth are then set individually to the required angle to give the saw blade clearance in its kerf. This is easiest with a tool called a saw set, which is used to bend the tips of the teeth to the correct angle.

The tool is operated by pre-selecting the required points per 25 mm for the saw blade being sharpened. Starting at one end, the tool is then positioned over a tooth and the handles squeezed. This will automatically set the angle of the tooth tip correctly. The procedure is repeated for each alternate tooth and then repeated from the opposite side for the teeth angled in the other direction.

Sharpening puts a cutting edge on the teeth. It is completed with a triangular file, holding the file at an angle to suit the type of saw blade and lightly filing each alternative 'V' between the teeth three times. When arriving at the end of the saw the blade is turned through 180° and the process repeated for the other teeth.

Figure 5.7 Saw blade held in saw stocks

Figure 5.8 Topping a saw blade

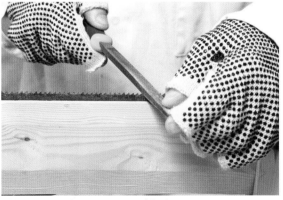

Figure 5.9 Shaping a saw blade

Figure 5.10 Setting saw teeth

Figure 5.11 Saw set tool

Figure 5.12 File used for sharpening saw teeth

Figure 5.14 Sharpening a twist bit with a triangular file

Figure 5.15 Sharpening a twist bit with a flat file

Rip saw teeth are filed at right angles across the blade to create a chisel effect. Cross cut saw teeth are normally filed with the file at approximately 70°, sharpening both the front of one tooth and the back of the next at the same time.

Sharpening twist bits

The life of a twist bit is shortened every time it is sharpened, so only sharpen when necessary. Twist bits can come with or without spurs. See Figure 5.13 for a twist bit with spurs. The spurs are the first to become dull, followed by the cutters. If the screw point becomes blunt the twist bit must be replaced.

Twist bits can be sharpened with various shaped files, the shape depending on the type of twist bit.

Tang

Shank

Spur

Cutting edge

Feed screw

Cutting edge

Twist

Spur

Figure 5.13 Parts of a twist bit

Figure 5.16 Saw teeth covered for protection

Tool protection

When stored in a tool bag, tool box or cupboard all tools should be prevented from moving about, otherwise cutting edges and sharp points can be damaged. Some useful tips for protecting tools are given below.

- Planes should have blades retracted before being stored.
- Protective covers can be made or purchased for saw teeth.
- Most chisels are provided with plastic end covers when purchased and should be used.

A strong, purpose-made chisel roll is also useful when storing in a tool bag or box.

- The sharp points on twist bits should always be protected as, unlike chisels, reshaping is difficult.

K2. Know how to use handsaws

There is a wide range of tools that cut using saw teeth. They can be classified as:

- handsaws, including rip saw, cross cut and panel saw
- back saws, so named because they have reinforcing metal along the back edge, including tenon, dovetail and bead saw
- saws for cutting circles or curves, including bow saw, coping saw, fret saw, pad, compass or keyhole saw.

They vary in the way they cut depending on the shape and size of the teeth, also on the angle the teeth are bent outwards from the line of the saw blade. This angle is called the set of the teeth. Cross cut teeth are designed to act like tiny knives that sever wood fibres while rip saw teeth are shaped to act like small chisels.

Saws are still commonly categorised by the number of 'teeth per inch' (TPI), but this is now taken to mean the number of teeth every 25 mm, as we have used below.

Many modern saws are designed to be used until blunt and then replaced. On older saws teeth need to be regularly maintained, which involves sharpening and setting.

Rip saw

Rip saws are usually used for cutting with the grain. Typically they are 650 mm to 700 mm long with 3–4 teeth per 25 mm. The teeth are filed at 90° across the blade, shaped like chisels and thus may be described as a gang of cutting chisels in a row. The saw should be at 60° to the work to cut most efficiently.

When starting to cut with a saw, make the first cut by drawing the saw backwards, as this avoids the saw jumping out from the mark, which can cause injury. Use your thumb as a guide as you start to cut.

Cross cut saw

Cross cut saws, as the name implies, are used for cutting across the grain, but can also be used for short rips on light timber.

Typically they are 650 mm long with 5–8 teeth per 25 mm. The teeth are bevelled (i.e. filed at an angle) across the saw to produce

Figure 5.17 Chisel roll

Figure 5.18 Chisel with plastic end cover

Did you know?

The cut that the saw makes is called a saw kerf

Figure 5.19 Cross cut saw

Figure 5.20 Panel saw

Figure 5.21 Tenon saw

a knife-like edge on the forward and back edge of the tooth. The angle of this bevel is between 60° and 75°.

The cutting edge should ideally be at 45° to the work when sawing across the grain.

Panel saw

Panel saws are used for cutting plywood, large tenons and most fine work (e.g. on polished materials). They are obtainable up to 600 mm long. The teeth are practically the same as the cross cut saw, but usually there are 7–12 teeth per 25 mm.

Tenon saw

The tenon saw has teeth in a similar pattern to a cross cut saw but with 12–14 teeth per 25 mm. It is used for cutting joints and general bench work. There is a reinforcing strip along the top of the blade made from steel or brass to keep the saw rigid. Tenon saws are typically 300 mm to 350 mm long.

There is no special virtue in brass or steel for the reinforcement, except that brass is kept clean more easily.

Dovetail saw

The dovetail saw is a smaller edition of the tenon saw with the same tooth pattern, but the teeth are finer, 18–24 per 25 mm, and it has a thinner blade. It is used for dovetails and other fine work. Dovetail saws are obtainable up to 200 mm in length.

Bead saw

Even finer than the dovetail saw is the bead saw or 'gents saw' as it is sometimes known. They have 15–25 teeth per 25 mm and are designed for very delicate work. They are usually 150 mm to 200 mm long.

Bow saw

The bow saw, also known as a frame saw because of its construction, has a thin blade and is used for cutting circular or curved work. The blade is held in tension by a cord twisted tourniquet fashion, though modern versions are fitted with an adjustable steel rod. The teeth are of the cross cut saw pattern and blades are 200 mm to 400 mm in length.

Figure 5.22 Bow saw

Coping saw

The coping saw has a very narrow blade held in tension by the springing of the frame and is also used for cutting curves, especially internal and external shapes. The teeth are of the rip saw pattern, typically 14 teeth per 25 mm and blades can be 150 mm long.

For internal shapes the blade can be released from the frame and reconnected and tensioned after pushing the blade through a hole.

Figure 5.23 Coping saw

Fret saw

Fret saws are very similar to coping saws but have a deep throat in the frame and smaller, finer blades, up to 32 teeth per 25 mm. They are used for cutting very tight curves and blades are generally no more than 125 mm long.

Pad, compass or key hole saws

Pad saws have no frame, so they can be used where coping or fret saws cannot reach. They tend to have a single handle with interchangeable blades to carry out a range of tasks, especially for cutting key holes and large internal shaped work. The handle may be angled to the blade. Key hole saw handles are usually in line with the blade.

Blades are tapered with a slot, which fits into the saw handle and is then secured with two screws. Blades for cutting wood may come in lengths from 125 mm to 375 mm and have teeth set in the rip saw or cross cut saw pattern. Other blades will cut plastic, metal, etc.

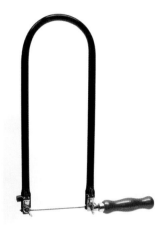

Figure 5.24 Fret saw

Figure 5.25 Pad saw

Hack saw

A hack saw is a framed saw that is used to cut metal components such as pipes. A hack saw can cut through copper, brass and steel.

Figure 5.26 Key hole saw

> **Remember**
>
> Most modern saws require little maintenance, but should have teeth covered for protection when not in use and be wiped down with a clean, lightly oiled rag

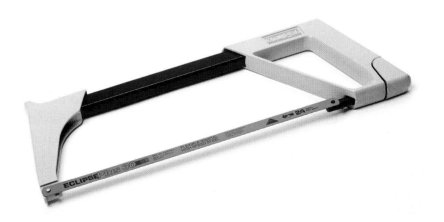

Figure 5.27 Hack saw

Figure 5.28 Flooring saw

Figure 5.29 Mitre box

Figure 5.30 Mitre block

Figure 5.31 Bench hook

Flooring saw

Flooring saws are specially designed to cut through floorboards in situ. They have a curved blade, which means they can cut into a board without having to drill through first, and also cause less damage to neighbouring boards. There are also teeth on an angled front edge, which allow cutting into skirting boards.

Blades are short and stiff, generally no more than 320 mm, with 8 teeth per 25 mm.

Mitre boxes and blocks

Mitre boxes and blocks are used to guide the saw when cutting angles, particularly mitres. Usually they are made to suit the job and replaced when worn, though purpose-built versions can be purchased.

Bench hook

A bench hook is a very simple device for holding the timber on a bench when sawing.

Hazards associated with handsaw use

Sawing is one of the most basic but essential skills for a carpenter or joiner. As with using any tool, safety is paramount and you must ensure that you have secured your workpiece and are wearing the appropriate PPE for the job at hand.

Once the area has been marked out for the cut, place the saw onto the work and – ensuring your arm is straight and your other hand is away from the blade – slowly pull back on the saw using little pressure. Repeat a couple of times until a groove is formed.

You can now start to push the saw through the groove, using it to cut on the push stroke rather than the pull. Do not apply too much pressure or force the saw, and keep your arm straight and in line. Failure to keep your other hand at a safe distance from the workpiece may result in the saw coming free from the saw cut and cutting into your hand.

K3. Know how to use hand-held planes

Planes and chisels

New plane or chisel blades will come with a bevel of 25° on the cutting edge. This has to be sharpened (honed) to an angle of 30° before use and regularly re-sharpened to this angle. See Figure 5.1 on page 185.

Planing tools

Planing tools are used to cut thin layers of wood, leaving a flat surface. They come in many forms, each developed to carry out a specific function or job, from levelling a surface to cutting bevels, rebates or grooves. The most widely used are as follows:

- smoothing plane
- jack plane
- jointer or trying plane
- rebate plane
- plough plane
- shoulder plane
- bull nose plane
- block plane.

The first three are often referred to as bench planes. The different parts of a smoothing plane are labelled in the photo. Other planes have similar component parts.

Figure 5.32 Smoothing plane

Labels: Handle, Cutting iron, Cam, Cap iron, Cap iron screw, Knob screw and nut, Heel, Knob, Adjusting nut, Toe, Steel body

Did you know?

Chamfering is to bevel an edge or corner, usually cutting an equal amount from each face

Planes are available in a variety of lengths as they are used for different tasks. For example, you wouldn't use a bull nose or block plane to plane a door as the base of the plane is not long enough and you may end up planing the door edge concave. A jack plane should be used for this job, but shouldn't be used to clean up the mitres on architrave. Ensure you select the right plane for the task at hand.

Smoothing plane

The smoothing plane is the shortest of the bench planes and is used for final finishing or cleaning up, bevelling and chamfering. It can be used to follow the grain and can even be used with one hand.

Jack plane

The jack plane is a medium length bench plane, generally 350 mm to 380 mm, with similar blade width to the smoothing plane. It is used for rapid and accurate removal of waste material such as dressing doors for hanging.

Jointer or trying plane

The jointer, try or trying plane (sometimes also known as the long plane) is the longest bench plane, ranging from 450 mm to 600 mm. It is generally used for smoothing long edges of timber. The length helps to make the surface as level as possible. Typical blade width is 60 mm.

The sequence of operations for squaring off a length of timber is:

1. One surface is planed flat.
2. Then plane one edge square to this dressed face.
3. The wood can then be dressed on all round to the required width and thickness.

Figure 5.33 Jack plane

Remember

When bevelling or chamfering, mark with a pencil not a gauge because the gauge marks will still show when the chamfer or bevel is planed

Figure 5.34 Jointer plane

Rebate plane

Rebate planes, as the name suggests, are used for making and cleaning out rebates. The simplest is known as a bench rebate plane. It is similar to the other bench planes described above but has the cutting blade extending across the full width of the sole and at an angle of 55° to 60°. It is usually 37 mm wide and at right angles across the sole, but can be obtained with the blade set at a skew.

More complex versions are called 'rebate and fillister planes', which have removable guide fences and depth gauges to assist in making accurate cuts. This avoids having to clamp a batten to the wood to act as a guide, which is advised with a standard bench rebate plane.

Plough plane or combination plane

Plough planes also have guide fences and depth gauges. They are usually used for cutting grooves along the grain of the timber, as well as rebates, typically from 3 mm to 12 mm wide. However, they come with a multiple choice of square and shaped cutters adapted for shaping **tongues**, **bends**, **beads**, **mouldings**, etc. The combination plane is essentially the same but may have additional capabilities.

Shoulder plane

Shoulder planes are similar to rebate planes but, as the name implies, they are primarily for shaping the shoulders of tenons, etc. They are often much narrower with blade widths down to 15 mm.

Bull nose plane

Bull nose planes are similar to shoulder planes but have the cutting edge very close to the front of the plane. Hence, they are particularly useful for cleaning out the corners of stopped housings or rebates. Some have a removable front end so that the plane can be worked right into the corners. They are obtainable with cutter widths from 10 mm to 28 mm.

Figure 5.35 Rebate plane

Key terms

Tongues – the projecting pieces of timber that fit into a groove in a tongue and groove joint

Beads – the shaped pieces of timber added to the end of other pieces of timber to give a desired finish

Mouldings – decorative finishes around door openings and at floor and wall junctions

Figure 5.36 Plough plane

Figure 5.37 Shoulder plane

Figure 5.38 Bull nose plane

Figure 5.39 Block plane

Did you know?

Block planes are called this because they were used to cut butcher's blocks, which traditionally used end grain timber

Figure 5.40 Round face spokeshave

Figure 5.41 Flat face spokeshave

Compass plane

Compass planes have a flexible sole to enable them to be used on concave or convex surfaces. There is screw adjustment to change the radius of the curve, within limits. Blade width is typically about 45 mm.

Block plane

Block planes come in a variety of designs but their key difference from other planes is the low angle at which the cutting blade is set, typically 20° but can be as low as 12°. This enables them to cut end grain.

Shaping tools

Even though work on curved surfaces can be carried out with planes and chisels, the following tools are used for shaping curved surfaces and edges:

- spokeshaves
- rasps and files
- Surforms®.

Spokeshave

Spokeshaves perform a similar role to a smoothing plane, except that they can do this on curved surfaces. They have a steel cutter set in either wood or metal handles and are designed to be used with two hands, pushing the tool along the wood to remove shavings.

There are two main types, a round face for concave surfaces and a flat face for convex surfaces. Cutter widths are typically around 50 mm, but they can be up to 100 mm.

Figure 5.42 Files

Rasps and files

Rasps and files have a hard, rough surface of miniature teeth designed to smooth wood or metal by removing very small shavings. Rasps have teeth formed individually, whereas files have teeth formed by cutting grooves into the surface in patterns.

Files, in particular, come in a wide range of shapes and sizes, with rectangular, triangular, oval or circular cross-sections, and combinations of these. They are generally classified by the number of teeth per 25 mm and the pattern of cuts used to create them. These may range from 26 teeth per 25 mm on coarse files to 60 teeth on smooth ones, and the grade may differ between the faces on the same file. Generally the longer the file, the coarser it is.

Rasps are mainly flat, round or half round and the teeth coarse. They are mainly used for the initial shaping of wood, able to remove a lot of material quickly but leaving a rough surface which then needs to be smoothed with a file or spokeshave.

Figure 5.43 Rasp

Surforms®

Surforms® differ from rasps because the teeth are punched right through the metal, enabling timber shavings to pass through. Hence, the tool is less inclined to clog up. They can be flat or round, the latter being useful for enlarging holes and shaping cuts.

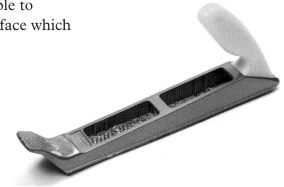

Figure 5.44 Surform®

K4. Know how to use hand-held drills

Drilling or boring tools

There are countless reasons why holes may be required in timber, for example to insert screws, dowels or other fittings. Hand tools used for these tasks can be broken down into:

- bradawls
- hand drills
- carpenter's brace.

Many of these tasks, traditionally completed with hand tools, have been taken over by the power drill. In particular, the hand drill and carpenter's brace are much less used today. However, hand tools for making holes still play an important part in the workplace and a good carpenter and joiner should be able to use them all.

Find out

Is it possible to repair damaged timber? If so, what methods can be used to fix a split, a minor gauge, or warped wood?

It is vital to know how to operate hand-held drills correctly. Improper drilling methods can damage the timber you are working with, adding delays to the job and costing your company money.

Using excessive force or pressure when drilling can cause cracks or splits in the timber. Failure to clamp the wood when drilling through could lead to splintering and other damage as the last piece of wood is forced through the hole by the drill. Lack of attention when drilling could lead to gouges in the wood if your tool slips.

Bradawl

The bradawl is used for making pilot holes for small screws or a centre mark for drilling. Some have sharp points, others have tips like a sharpened screwdriver blade or even spiral cutters. The points work by pushing aside the wood fibres when pressure is applied, but this can cause the wood to split. A sharp screwdriver tip can reduce this risk by cutting the fibres initially, the hole then being further deepened by a point.

Hand drill or wheel brace

Hand drills, or wheel braces, are useful for boring screw holes up to 6 mm in diameter. A selection of twist bits is required, usually with a round shank as used with power drills.

Carpenter's brace

The carpenter's brace has been in use for several hundred years and can hold a wide variety of bits for boring holes and other tasks. Most types of brace have a ratchet mechanism so that the tool can be used in restricted spaces.

The chuck of the brace is designed to grip bits with square shanks, called a tang. However, some braces can accept rounded shanks that are designed for power tools. The main types of bits are described below and on the page opposite.

Centre bit

A centre bit is a short, fast cutting bit. It has a lead screw and a single cutting spur. It is used to bore accurate but shallow holes. To go right through the wood it is normal to drill into the wood until the point comes through, then use this as the centre mark to drill from the other side.

> **Did you know?**
>
> You can use a wire brush to clean rasps and files when they become clogged with shavings

Figure 5.45 Bradawl

Figure 5.46 Hand drill

Figure 5.47 Carpenter's brace

Figure 5.48 Irwin bit

Twist bit

Twist bits have a lead screw and two cutting spurs. The shank has a helical twist or spiral that clears the wood from the hole. Irwin bits have a single spiral, the Jennings a double spiral. Both are used for boring holes from 4 mm upwards. The Jennings bits are claimed to be more accurate and stronger because of the double spiral, so are favoured for longer bits, but, in practice, there is little to choose between them.

Figure 5.49 Jennings bit

Forstner bit

Forstner bits are used for high-quality, accurate work where a flat bottom to a hole is required. They have no threaded point and need a pilot hole made by other cutters to start and then steady pressure to make them cut. They can range in size from 10 mm to 50 mm.

Expansion bit

Expansion bits are similar to centre bits but with an adjustable cutting spur. They are only designed to cut a shallow hole.

Figure 5.50 Forstner bits

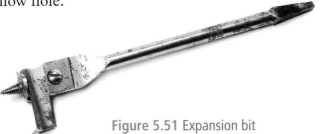

Figure 5.51 Expansion bit

> **Safety tip**
>
> Never use an expansion bit in any power tool

Countersink bit

Countersink bits are used to form a recess for a screw head. They come with different cutter shapes. Snail countersinks are used in braces, and rose countersinks in hand drills or power drills.

Figure 5.52 Snail countersink

Screwdriver bit

A carpenter's brace with a screwdriver bit can be very useful for removing stubborn and worn screws, as considerable pressure can be applied and the brace handle gives excellent leverage. They are often more controllable than a power drill and less likely to jump out, which can damage the screw further. They can also be used to insert screws.

Figure 5.53 Rose countersink

Figure 5.54 Screwdriver bits

Lip bit

Also known as a lip, wood or dowel bit, spur point bits have a central point and two raised spurs that help keep the bit drilling straight. The bit cuts timber very fast when used in a power drill and leaves a clean sided hole. They are ideal for drilling holes for dowels as the sides of the holes are clean and parallel.

Auger bit

Auger bits are generally used in hand drills, and are designed to drill wide and deep holes. Holes cut by an auger bit will have a flat bottom. Because the centre of the bit bites into the wood and pulls the drill into the timber, auger bits are unsuitable for use with a power drill.

Brace bit

Brace bits are less common in modern woodwork, but would be used with a carpenter's brace (see page 192).

Masonry bit

Masonry bits are designed for drilling into brick, block, stone, quarry tiles or concrete. The cutting tip is often made from tungsten carbide bonded to a spiralled steel shaft.

Flat wood bit

Intended for power drill/cordless use only, the centre point locates the bit and the flat steel on either side cuts away the timber. Flat wood bits are used to drill fairly large holes. As they give a flat bottomed hole with a central point they are ideal when the head of a screw or bolt needs to be recessed into the timber.

Holding and clamping tools

Joiners frequently need to grip items and hold them steady while using tools, or while fixing them in position with nails, screws or glue. These include simple tools like pliers and pincers, as well as carpenters' vices attached to fixed or portable work benches, and a range of clamps (or cramps).

Pliers

Pliers are a universal gripping device used to grip small nuts and bolts. They can also be used to bend and cut, as well as to strip wire.

Pincers

Pincers are the simplest gripping tool and are an essential wood-worker's tool. They are primarily used for removing nails but

Figure 5.55 Pliers

have many other uses. A good-quality pair should give many years of service.

When using pincers for removing nails they should be operated similarly to a claw hammer. Scraps of wood should be used to avoid damage to the timber and thicker pieces used to get additional leverage. There is often a small claw on one handle, which is useful for removing tacks.

Figure 5.56 Pincers

Clamps

Clamps, or cramps, have been developed to hold almost any shape of wood or other material in almost any position. Some are spring-operated, others have screw threads to pull jaws onto the material, others may use web straps that can be ratcheted tight, often around a frame. They may incorporate quick-release mechanisms.

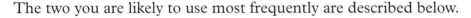

The two you are likely to use most frequently are described below.

Sash clamp

Sash clamps are designed to hold large pieces of wood or frames together, usually while gluing.

A steel bar, either flat or T-section, has holes drilled along its length. An end shoe, flat to fit over the edge of the work to be clamped, is able to be moved a limited distance down the bar by using a screw mechanism driven by a tommy bar. A second shoe, the tail shoe, is designed to fit over the other edge of the work. It can be locked in place by a steel pin through any of the holes in the bar.

With the end shoe withdrawn towards the end, the clamp is placed over the work and the tail shoe moved to give a loose fit. It is then locked in place with the pin. Pressure is then applied by screwing down the end shoe. The clamp must be square across the job to get even pressure.

Did you know?

Wooden packers can be used on shoes to avoid damage to the work

Figure 5.57 Sash clamp

Figure 5.58 G clamp

Figure 5.59 Retractable knife

G clamp

G clamps take their name from their shape, which resembles the letter 'G'. They come in various sizes and forms, and may include a quick-release mechanism but they all have a fixed end and an adjustable end.

They are very versatile and particularly useful for clamping work pieces to the bench to allow operations to be carried out such as routing or sawing. Again, pieces of scrap material can be used as packers to prevent damage to the work.

K5. Know how to use woodworking chisels

There is a wide range of tools designed to cut material using a sharp blade rather than saw teeth, including knives, scissors, chisels, gouges and axes. Chisels, gouges and axes are covered in more detail below.

Chisels

As with saws, chisels are available in a variety of shapes and sizes. The ones that you are most likely to use are described below. Like all edge tools, chisels work best when they are very sharp. This ensures less effort needed to cut, which enables greater accuracy.

With their sharper cutting edges, chisels can be a dangerous implement to work with if not used properly. Make sure that you carry chisels by the handle, and when passing a chisel to a colleague, safely hold the blade while offering the handle.

Firmer chisel

A firmer chisel is one of the strongest chisels. It is used for general purpose wood cutting and designed to be used with a mallet, if required.

Safety tip

Never hit a chisel with your hand as this may cause injury

The blade is generally about 100 mm long when new, rectangular in cross-section and tapering slightly from the bolster to the cutting edge. Size is normally by blade width, which may range from 6 mm to 50 mm.

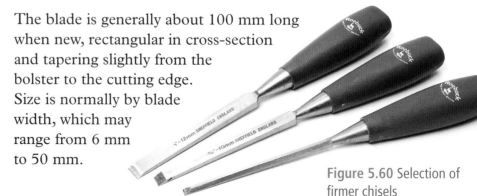

Figure 5.60 Selection of firmer chisels

Bevelled-edge chisel

Bevelled-edge chisels are variations on the standard firmer chisel and come in similar sizes. The two long edges are bevelled, which makes them lighter but not as strong. Hence, they should not be used with a mallet, except for very light taps.

They are used for short paring and other fine work. The bevelled edge also helps when cleaning out corners that are less than 90°.

Figure 5.61 Bevelled-edge chisels

Paring chisel

Paring chisels may be rectangular in cross-section or have bevelled edges. They are longer, normally around 175 mm, more slender than the firmer chisels and are, thus, a more delicate tool. Their main use is for cutting deep grooves and long housings. They should not be hit with a mallet.

Mortise chisel

The mortise chisel is designed for heavy-duty work, with a thick, stiff blade and generally shorter than the firmer and paring chisel. It may also have a slight bevel on all edges to allow easy withdrawal from the work. They are typically 6 mm to 50 mm wide.

Figure 5.62 Paring chisel

The handle is longer than on other chisels, with a wide curved end designed to take blows from a mallet. It is sometimes encircled with a metal ring to give extra strength and a leather washer inserted between the shoulder of the blade and the handle to absorb mallet blows. The handles can either be held by a simple tang or in a socket.

Variants of the heavy-duty mortise chisel are the sash mortise chisel for lighter work and the lock mortise chisel, with a swan neck, useful for removing waste from deep mortises.

Figure 5.63 Mortise chisel

Gouges

Gouges are a type of chisel with a curved cross-section to the blade. They come in two main types, similar to chisels, called firmer and paring.

Firmer gouge

Firmer gouges come with two blade types: out-cannel and in-cannel.

- Out-cannel gouges have the cutting edge ground on the outside, so that they can be used to make concave cuts (i.e. cuts into the surface).

- In-cannel gouges have the cutting edge ground on the inside, so that they can be used to make convex cuts (i.e. leaving a protruding bulge on the surface).

Blade length is about 100 mm when new and they are designed to be used with a mallet. Blade widths typically range from 6 mm to 25 mm, though larger ones are available. Both are usually ground square across, though the out-cannel blade is sometimes rounded for cutting deep hollows.

Figures 5.64 and 5.65 show the two gouges being used to make convex and concave cuts.

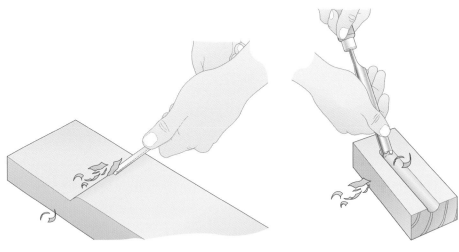

Figure 5.64 In-cannel gouge in use

Figure 5.65 Out-cannel gouge in use

Paring gouge

Similar to paring chisels, and sometimes known as scribing gouges, paring gouges are longer and thinner than firmer gouges and used for finer work. They are only ground on the inside for in-cannel cuts. They are not designed to be used with a mallet and may have a double bend in the neck to keep the handle clear of the work surface.

New planes and chisels

New plane or chisel blades will come with a bevel of 25° on the cutting edge. This has to be sharpened (honed) to an angle of 30° before use and regularly re-sharpened to this angle. See Figure 5.65.

Axes

When an axe is used correctly, it is highly effective at removing waste wood and cutting wedges. However, it must be kept sharp, and the edge protected when not in use.

Axes vary in the shape of the head and its weight. The head is fitted on to a wooden shaft, preferably hickory, and held by wedges in a similar manner to a hammer. Axes designed to be used with one hand are typically around 1 kg in weight. Felling axes, with longer handles and which are designed to be used with two hands, may be 2.75 kg or heavier.

Figure 5.66 Hand axe

Working Life

On his first day on site, Julio went for lunch having left his tools in a pile near the entrance. While he was on lunch, it began to rain. Julio was surprised to be reprimanded by his supervisor when he got back from lunch.

What should Julio have done with his tools? Julio must always make sure that his tools are in the best possible condition so he will need to take care of them both when he is working with them and when he leaves them somewhere.

What risks are there in leaving tools outside and near the entrance to a site? There might not just be safety risks, but also other possible hazards or dangers.

Functional skills

At the end of this unit you will have the opportunity to answer a series of questions on the material you have learnt. By answering these questions you will be practising the following functional skills:

FE 1.2.3 Read different texts and take appropriate action.

FE 1.3.1 – 1.3.5 Write clearly with a level of detail to suit the purpose.

FM 1.1.1 Identify and select mathematical procedures.

FM 1.2.1c Draw shapes.

FAQ

Do I have to buy every type of plane there is?

No, most carpenters and joiners will have a selection of planes. The work you do – whether it is site-based or in a joiner's workshop – will determine the tools you require. For example, a site carpenter can get by with a block plane and a smoothing or jack plane.

Do I have to use a mortise chisel for chiselling out mortises?

No, bevel edge or firmer chisels will also do the job. However, care must be taken when using bevel edged chisels as they are not as strong.

Why would I need a jig?

Jigs are vital in maintaining accuracy, especially if you have more than one component to work.

Check it out

1   Name three types of rules which could be used by carpenters and joiners, stating the advantages of each.

2   What are the following used for:

  a)   sliding bevel?

  b)   tri-square?

  c)   combination square?

3   State the uses of the following:

  a)   smoothing plane.

  b)   rebate plane.

  c)   plough plane.

4   Name four types of bit that can be used with a ratchet brace.

5   Name and state the uses of three different types of saw.

6   Sketch the following:

  a)   bevelled edge chisel.

  b)   mortise chisel.

  c)   paring chisel.

7   Explain what the Kitemark and double square symbol indicate.

8   List two things that should be done to help with tool protection.

9   Show with sketches the difference in toothing between a tenon saw and a rip saw.

10   Sketch the difference between an in-cannel and an out cannel-gouge.

Getting ready for assessment

The information contained in this chapter, as well as continued practical assignments that you will carry out in your college or training centre, will help you with preparing for both your end of unit test and the diploma multiple-choice test. It will also aid you in preparing for the work that is required for the synoptic practical assignments.

The information contained within this chapter will aid you in learning how to identify various hand tools as well as giving you information on their safe uses.

You will need to be familiar with how to:

- maintain and store hand tools
- use hand-held drills
- use handsaws
- use woodworking chisels
- use hand-held planes

This unit should have made you familiar with the types of tools you will encounter in carpentry, and the purposes of each of these tools. You should also be familiar now with how to use these tools. These skills will be vital for all carpentry work, not only in the synoptic practical assignments, but also in your professional career.

For example, learning outcome 2 requires you to be able to use a handsaw. This unit has shown you the different types of handsaw and their purposes, as well as the correct method to be used and the hazards caused by using a handsaw incorrectly. You have also seen why different sized saw teeth have different uses. For any test or job, you will need to make sure that you have set up and are using a tool correctly, otherwise you may cause harm not only to the thing you are working on, but also yourself!

You will need to use the knowledge you have gained about handsaws to carry out your own practical work. This could include cutting across the grain, with the grain and any curves. Remember you will always need to make sure that you are cutting within any specification.

The speed and accuracy with which you use these tools will also help you to prepare for the time limit and the quality of work that will be required to pass the synoptic practical task. But remember, do not try to rush the job as speed will come with practice and it is important that you get the quality of workmanship right. You will also need to be sure that you are using the tools safely.

Before you start work on the synoptic practical test it is important that you have had sufficient practice and that you feel that you are capable of passing. On approaching the test it is best to have a plan of action and a work method that will help you as well as a copy of the required standards, associated drawings and sufficient tools and materials. It is also wise to check your work at regular intervals. This will help you to be sure that you are working correctly, and help you to avoid problems developing as you work.

Always make sure that you are working safely throughout the test. Make sure you are working to all the safety requirements given throughout the test and wear all appropriate personal protective equipment. When using tools, make sure you are using them correctly and safely.

Good luck!

CHECK YOUR KNOWLEDGE

1 Which tool could you use instead of a pencil when you want to mark measurements very accurately?
- **a** marking gauge.
- **b** tri-square.
- **c** marking knife.
- **d** callipers.

2 What are dividers used for?
- **a** accurate checking of widths and gaps.
- **b** marking out measurements.
- **c** measuring angles.
- **d** marking length and thickness.

3 Which saw is between 650 mm and 700 mm long, with 3–4 teeth per inch?
- **a** cross cut saw.
- **b** rip saw.
- **c** tenon saw.
- **d** coping saw.

4 The best saw for cutting curved work is a:
- **a** bow saw.
- **b** panel saw.
- **c** coping saw.
- **d** bead saw.

5 Which saw is best for cutting metal or plastic?
- **a** flooring saw.
- **b** fret saw.
- **c** pad saw.
- **d** hack saw.

6 Which chisel is best for convex and concave work?
- **a** paring chisel.
- **b** firmer chisel.
- **c** gouge chisel.
- **d** bevel-edge chisel.

7 Which plane is best for clearing the corners of stopped housings?
- **a** rebate plane.
- **b** bull nose plane.
- **c** plough plane.
- **d** block plane.

8 Which hand tool is best for making small pilot holes?
- **a** rasp.
- **b** bradawl.
- **c** punch.
- **d** countersink bit.

9 The square shank of a drill bit is called a:
- **a** ting.
- **b** tang.
- **c** tong.
- **d** tung.

10 Which screwdriver gives the best grip?
- **a** slotted head.
- **b** Phillips crosshead.
- **c** Posidrive.
- **d** all of the above.

11 The most common levelling tool is the:
- **a** spirit level.
- **b** laser level.
- **c** water level.
- **d** plumb bob.

12 The correct sequence for sharpening a saw is:
- **a** topping, cutting, setting, sharpening.
- **b** cutting, topping, setting, sharpening.
- **c** topping, setting, cutting, sharpening.
- **d** topping, setting, sharpening, cutting.

13 The grinding angle of planes and chisels is set at:
- **a** 20°.
- **b** 25°.
- **c** 30°.
- **d** 35°.

14 When sharpening a chisel, why is it important to use all of the oilstone?
- **a** to prevent the stone clogging.
- **b** to prevent the chisel breaking.
- **c** to avoid hollowing the stone.
- **d** to prevent a wire edge forming.

Unit 1006

Maintain and use carpentry and joinery power tools

Whatever type of work is involved, the quality of the finished product depends on the skill of the operator in selecting and using the correct tools to cut, shape and assemble the materials for the task. There is a wide range of tools available and this chapter looks at the main power tools that the carpenter and joiner should be able to use and maintain properly. We will also look at the importance of safety and the precautions you should take when using tools. Issues relating to the protection and maintenance of tools will also be addressed.

This unit also contains material that supports TAP Unit 3 Install Second Fixing Components. It also contains material that supports the delivery of the five generic units.

This unit will cover the following learning outcomes:

- How to maintain portable power tools
- How to use portable power tools to cut, sand and finish
- How to use portable power tools to drill and insert fastenings

Functional skills

When reading and understanding the text in this unit, you are practising several functional skills.

FE 1.2.1 - Identifying how the main points and ideas are organised in different texts.

FE 1.2.2 – Understanding different texts in detail.

FE 1.2.3 – Read different texts and take appropriate action, e.g. respond to advice/instructions.

If there are any words or phrases you do not understand, use a dictionary, look them up using the internet or discuss with your tutor.

Remember

You must receive training and obtain permission before using any power tool and an instructor must be present while you carry out the operation

Functional skills

When using power tools you will need to make sure that you are working safely. To do this you will need to practise **FE** 1.2.1 – 1.2.3 which relate to reading and understanding information including different texts and taking appropriate action, e.g. responding to advice/instructions.

K1. Know how to maintain portable power tools

By the end of this section you will have the basic knowledge to begin to use portable powered hand tools safely but, as with any tool, you will need practice to become competent in their use.

These tools are covered by legislation and you must fully comply with the following:

- **The Provision and Use of Work Equipment Regulations 1998** – The regulations that deal with all power tools
- **The Abrasive Wheels Regulations 1974** – the regulations that deal with tools that have an abrasive wheel such as a grinder
- **The Protection of Eye Regulations 1974** – the regulations that deal with protection of eyes
- **The Health and Safety at Work Act 1974** – the regulations that deal with all work that is carried out.

Summary of power tool types

The following power tools are described in this unit:

- power drills
- powered planer
- routers
- portable circular saws
- chop saws
- jigsaws
- sanders
- cordless tools.

First we will cover:

- safety issues common to all power tools
- power supplies
- power tool maintenance.

Safety first

If used correctly, powered hand tools can save time, money and effort. Skill in using them comes only through training and experience – it cannot be picked up on site. All the safety precautions for hand tools apply, but there are additional risks with power tools, especially portable ones.

Functional skills

While working through this unit, you will be practising the functional skills **FE** 1.2.1 – 1.2.3. These relate to reading and understanding information. As with the other sections in the book, if there are any words or phrases you do not understand, use a dictionary, look them up using the internet or discuss with your tutor.

Although each type of power tool has its own safe working procedures the following basic rules apply to *any* portable powered tool.

- Only use powered hand tools if you have been authorised to do so and have been taught how to use them correctly.
- Never use a power tool without permission.
- If you are unfamiliar with the equipment, read the instruction booklet and practise using the tool. Go ahead with the job only when you are sure you can do so safely and be certain of producing good results.
- Always select the correct tool for the work in hand. Check that it is in good condition and that any blade or cutter has been fitted correctly. *If in doubt, ask someone with experience.*
- Ensure that the tool and power supply are compatible. Do not mix voltages; for example 110 V tools must only be used on 110 V supplies.
- Ensure that plugs, wires, extension leads, etc. are in good condition. If not, do not use the tool.
- Ensure that any extension leads are fully uncoiled before use, as they can easily overheat.
- Never carry any tool by its lead.
- Always use fitted safety guards correctly, as these have been designed with your safety in mind. Never use a tool without them.
- Do not let others touch tools or extension cords while in use and disconnect tools when you are finished with them.
- Never put a tool down until all moving parts have stopped and, before making any adjustments, disconnect a tool from its power supply or switch it off at the source. Just switching off the tool is not good enough.
- Should a tool not operate, do not tamper with it but report the matter immediately to your supervisor. Repairs should only be carried out by a competent person.
- Wear goggles when there is any danger of flying particles. A dust mask or respirator, ear protectors and safety helmet may also be necessary.
- Report any accident (or 'near miss') to your supervisor, regardless of whether it results in injury to yourself or others.
- If you are injured, even when it is a minor wound or scratch, get immediate first aid treatment.
- Guard against electric shock. Hence, never allow electrical equipment to become damp or wet and never use electrical equipment in damp or wet conditions.

Safety tip

Tools cannot be careless but you can, so do the job safely – it is quicker in the long run

Figure 6.1 Extension lead on drum

Figure 6.2 Transformer used to reduce voltage from 230 V to 110 V

Safety tip

Just because safety information is displayed it does not mean there are no risks to consider. Tools must be correct for purpose and used in accordance with manufacturers' guidance

Power supplies

For tools that need to be connected to a power supply this can be 230 V (normal domestic mains supply) or 110 V (as a reduced voltage via a transformer). The use of tools at 230 V is not recommended, as an electrical shock from the tool at this voltage can be fatal. Only tools using a supply of 110 V, reduced from 230 V through a transformer, can be used on construction sites.

All power tools are now made with double insulation and should be stamped with a double insulation sign and a Kitemark. See Figures 6.3 and 6.4.

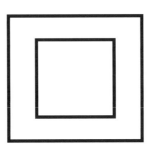

Figure 6.3 Double insulation sign

Figure 6.4 BSI Kitemark

Cordless tools

An increasing range of battery operated (i.e. cordless) tools is becoming available, including drills, jigsaws, sanders, screwdrivers, etc. They are particularly convenient in situations where the mains power supply is not easily accessible, or not available. There are no cables to cause problems when working some distance from a power point, or hanging down when working at height.

Cordless tools usually have the same chuck options and can carry out all the functions of tools connected to mains power. However, the heavier the task, the more rapidly batteries will discharge. Battery power is indicated by the voltage rating, for example 10 V, 12 V, etc. – higher voltage tools are more suited for heavy-duty work.

Batteries are rechargeable and this generally takes only a couple of hours. However, it pays to carry a spare, which can be used while waiting for the spent one to recharge. Battery life is considerably shortened if this is not done correctly, so always use the charger purchased with the tool.

When the battery is not connected to either tool or charger, ensure that nothing is allowed to touch the terminals – especially any metal – as this will cause a short and probably destroy the battery.

Figure 6.5 Cordless drill

Portable Appliance Test (PAT)

Portable appliance testing is a requirement of many regulations, including PUWER, which states 'that every employer shall ensure that work equipment is maintained in an efficient state, in efficient working order and in good repair'.

Tests must be carried out by a qualified person and will require the portable appliance to be plugged into a machine that checks the machine for earth, insulation or any other faults. A visual inspection will also take place, and if the item passes a sticker will be placed on the appliance and/or a certificate will be provided to say the equipment has been tested.

It is good practice for companies to keep a log of all equipment that has been/is to be PAT tested, as the test must be carried out on all portable electrical appliances – microwaves and toasters as well as power tools, so a company may have several items that need regular testing. A log of all equipment is the best way to keep a record of this.

PAT testing should be carried out at regular intervals, but the type of appliance and its uses will determine how often the testing takes place. High-risk and heavy-use appliances should be tested every 3 months, medium-risk appliances every 6 months and low-risk every 12 months. If an appliance is not tested within the required time frame and an accident or fault with the equipment causes an accident, the company will be liable and can be prosecuted under health and safety law. It is vital that all equipment is kept up to date.

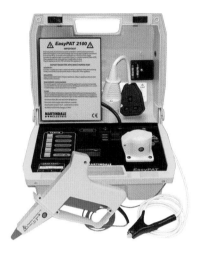

Figure 6.6 Portable electrical safety tester

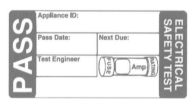

Figure 6.7 Electrical safety 'pass' sticker

Power tool maintenance

Most modern power tools are designed to operate over a long period of time with minimum maintenance. However, regular attention to powered hand tools improves safety and ensures that they work efficiently.

In particular they should be cleaned regularly (this is especially important with cutting and abrasive tools, because the dust produced can easily damage the motor). This may mean a daily clean for tools in constant use.

Information for cleaning and maintenance of power tools can be found in the instruction manual for each machine. This will clearly state which actions should and can be undertaken by you and which must be undertaken by authorised repair agents. Not only are repair agents trained, and equipped with the necessary specialist tools to complete the work correctly, but unauthorised repairs will invalidate any guarantee.

Regular visual checks must be made to spot damage to the tool, leads and plugs. Make sure that the ventilation slots are clear. Any damaged parts should be replaced before use, but only attempt this if it is a user-authorised task and you have been trained to carry it out.

Most portable power tools will come in a case. If not, the manufacturer's instruction manual often states how the tool should be stored. Proper storage is important, as poor storage will not only invalidate any guarantee but can lead to the tool being damaged which may make it unsafe.

K2. Know how to use portable power tools to cut, shape and finish

Power drills

Find out

What does double insulation mean?

Power drills are highly versatile and can be used with different bits, not just to drill holes in a range of materials but to drive or remove screws, cut circular holes with a saw cutter, or buff and polish. Various designs of chuck are available for rapid interchange of bits, the chuck size being matched to the power of the drill.

The simplest power drills operate at a single speed. However, most modern drills come with additional capabilities, which may include one or more of the following:

- dual speed, the slower speed for use when boring into brickwork or masonry using a tipped drill bit
- variable speed, controlled by trigger pressure
- reverse action, that is able to operate clockwise or anticlockwise and mainly used for driving or withdrawing screws
- percussion or hammer action, primarily for brickwork or masonry
- torque control, particularly useful to avoid over-tightening screws, nuts or bolts.

Choosing the most suitable drill and drill bit for the job depends on:

- the type of material that needs to be drilled
- size of holes required or task to be completed
- how often the drill will be used.

Figure 6.8 Power drill

Some of the common types of power drill are described below.

Palm grip drill

The most common and versatile of the electric drills, palm grip drills can be used to carry out work on timber, steel, alloy and masonry. Dust should be cleaned from bore holes at regular intervals.

They are available with single or twin speeds but may have other options as listed above.

Back handle drill

The back handle drill is a heavy-duty version of the palm grip drill. They have a larger chuck, allowing holes of greater diameter to be bored. They can also withstand long periods of heavy pressure but, as with the palm grip drill, dust should be cleaned from bore holes at regular intervals.

They are available in single-, two- and four-speed models but may also have other options as listed above. A handle can be attached to give greater control during use.

Rotary percussion (hammer) drills

Hammer drills have a percussive hammer effect which enables them to bore into hard materials. This percussive action is optional and is brought into operation by means of a switch. It is particularly effective for drilling into masonry or concrete. A depth gauge can be fitted to the side handle to provide accuracy where required.

Figure 6.9 110 V hammer drill

Good practice when using drills

- Always tighten the chuck securely.
- Take advantage of speed selection. Start drilling into hard materials at slow speed and increase gradually.
- Slow down the speed prior to breaking through the surface, particularly metals, to avoid snatching and twisting.
- Keep drill vents clear to maintain adequate ventilation.
- Use sharp drill bits at all times.
- Apply as much pressure as possible, consistent with the size of the drill. This is more important than speed, as it is essential to keep the edge of the drill biting into the material rather than let it rub on the bottom of the hole.
- Especially when using larger drill sizes, make a pilot hole with a smaller drill. For example, for a 13 mm-hole use a 10 mm or 12 mm drill and then open this out with a 13 mm drill.
- Check carbon brushes regularly. Excessive sparking indicates excessive wear or a possible short circuit. Report this to a supervisor.

- Make sure all plug connections are secure and correct.
- Keep all cables clear of the cutting area during use.
- Keep holes clear of dust. Drills should be withdrawn from holes at intervals, as an accumulation of dust not only causes overheating but also tends to blunt the drill bit.

Bad practice when using drills

- Never use a drill designed to operate with a 3-wire source (i.e. live, neutral and earth) on a 2-wire supply (i.e. live and neutral only). Always connect an earth.
- Do not use a drill bit with a bent spindle.
- Do not exceed the manufacturer's recommended maximum capacities for drill sizes on appropriate materials.
- Do not use high-speed steel (HSS) bits without cooling or lubrication.
- Never use a hole saw cutter without the pilot cutter.
- Do not cool the hot point of a drill by dipping it in water, as this will crack the tip.

Powered planers

Figure 6.10 110 V powered planer

Powered planers are invaluable for removing large amounts of waste wood, especially on site. With large models it is possible to remove up to 3 mm in one pass. The machine has an adjustable depth gauge to allow for the greater depths needed for rebating.

The sequence of operations is similar to using a hand plane. One surface is planed square and then one edge is planed square to this dressed face. The wood can then be dressed to the required width and thickness.

Remember:

- Check that the cutters are sharp to prevent overloading the motor and producing a poor finish.
- Secure material to be planed in a vice or on a workbench against a stop.
- Make adjustments to the planer before connecting to the power supply.
- Do not put the machine down until the cutters have stopped rotating.
- Wear ear, eye and nose/mouth protection.
- All machines should be checked on a regular basis by a qualified electrical engineer but the operator should check the visual condition of the planer, voltage, power cable and plug.

Routers

The router is a very versatile machine. Like a power drill, it has a chuck able to take different sizes of bit, the size depending on the power of the machine. However, the variety of bits and cutters that can be fitted is very wide and many different operations can be completed, including:

- cutting straight, curved and moulded grooves
- cutting slots and recesses
- rebating
- beading and moulding
- dovetailing
- laminate trimming.

Figure 6.11 110 V heavy-duty plunge router

Routers used by carpenters and joiners on site are most likely to be:

- heavy-duty routers
- heavy-duty plunge routers.

With the heavy-duty router the cutter projects from the base. Great care must be taken to feed it gradually into the material to prevent it from snatching.

Most routers now are of the plunge type, where the cutter is brought downwards into contact with the material to be cut and then withdrawn from the material when downward pressure is released. This is made possible by two spring-loaded plungers fitted to either side of the machine.

Both types are capable of producing excellent finishes, mainly due to the very high speed of the machine, but these speeds demand great control and concentration from the operator.

Figure 6.12 Router bits for grooving and trimming

For general work router bits and cutters are made of high-speed steel (HSS), but carbide-tipped bits and cutters offer a longer cutting life. Edging bits have a guide pin or roller which allows the cutter to follow an accurate path without biting into the material.

Routers come with various items as standard, including guides and fences, but some useful additional accessories are:

- trammel point and arm, used for cutting circles
- dovetail kit for producing dovetail joints
- bench or table stand that converts the router into a small spindle moulder
- trimming attachment to cut veneered edging.

Figure 6.13 Router bits for edging

Remember the following points.

- Secure the work piece before commencing work.
- Make sure the cutter/bit is free to rotate.
- Allow the machine to reach its maximum speed before making the first trial cut (into waste material).
- Make any cut from left to right.
- Move the router quickly enough to make a continuous cut but never overload the motor (listen to the drone) by pushing too fast, which will blunt the cutter and burn the wood.
- At the end of each operation, switch off the motor. The cutter/bit should be freed from the work and allowed to stop revolving before being left unattended.
- Goggles or safety glasses must always be worn, as well as other appropriate items of PPE.
- All machines should be checked on a regular basis by a qualified electrical engineer, but the operator should check the visual condition of the router, voltage, power cable and plug.

Portable circular saws

Portable circular saws are mainly used for cross cutting and ripping, but can also be used for bevel cuts, grooves and rebates. They can be used to cut a wide variety of materials:

- timber, including softwood and hardwood
- manufactured boards, including plywood, chipboard, blockboard, laminboard and fibreboard
- plastic laminates.

We will refer mainly to timber below, but the same principles apply to all materials.

Before use the saw should be adjusted so that, when cutting normally, the blade will only just break through the underside of the timber. This is achieved by releasing a locking device, which controls the movement of the base plate in relation to the amount of blade exposed.

A detachable fence is supplied for ripping. For bevel cutting, the base tilts up to 45° on a lockable quadrant arm. The telescopic saw guard, covering the exposed blade, will automatically spring back when the cut is complete.

Figure 6.14 110 V portable circular saw

Remember the following points.

- Wear appropriate PPE.
- Any timber being cut should be securely held, clamped or fixed, making sure that fixings are clear of the saw cut.
- Ensure the power cable is clear of the cutting action.
- Always use both hands on the saw handles, as this reduces the risk of the free hand making contact with the cutting edge of the blade.
- The saw should be allowed to reach maximum speed before starting to cut, and should not be stopped or restarted in the timber.
- At the finish of the cut, keep the saw suspended away from the body until the blade stops revolving.
- Disconnect the machine before making any adjustments or when not in use. It is not sufficient just to switch off the machine.
- Do not overload the saw by forcing it into the material.
- Keep saw blades sharp.
- Always work in safe, dry conditions.
- All machines should be checked on a regular basis by a qualified electrical engineer but the operator should check the visual condition of the saw, voltage, power cable and plug.

Chop saws

Chop saws have a circular blade rotating in a housing and a fixed bed onto which items to be cut can be fixed. The blade can be pulled down onto the work to cut it.

They can be fitted with different blades, depending on the material to be cut and the task. These can then be set to cut:

- square at 90° in both directions
- mitres at any angle up to 45°
- bevels up to 45°
- compound bevels up to 45° (i.e. two angles in one cut, both face side and face edge of the material).

Chop saws are capable of causing serious injury and should be securely fixed to a workbench or purpose-made stand at a height that is comfortable for the operator. The built-in bed tends to be short, so side extension tables or trestles must be used to support longer lengths of timber when cutting.

> **Remember**
>
> When cutting angles or bevels, the machine's effective cutting depth will be reduced

Figure 6.15 110 V chop saw

Remember the following points.

- Because of the high risk of injury when using this type of machine, safety must be a priority. **Read** and understand the manufacturer's instructions before using the machine.
- Chop saws are available for 230 V or 110 V, so be sure to connect the machine to the correct voltage.
- Ensure that cables are clear of the machine's moving parts, and are not causing obstruction to the materials to be cut, or in a position that may cause a person to trip.
- Check the machine for damage, wear and operating functions before use.
- Complete all machine settings (blade changes, angles, stops, guards, fences, etc.) when disconnected from the power supply. Blade changing must be carried out by a competent person and in conjunction with the manufacturer's instructions.
- Secure the work piece firmly to the machine bed. Hands should not be used to hold material in place while cutting.
- Concentrate on what you are doing. Keep an eye on the work being cut and *never* allow your hands to be closer to the blade than 150 mm.
- Do not force the machine to cut but allow it to cut at its own speed.
- Always use the recommended blade (blade size, tooth type, HSS, TCT, etc.). Do not use blunt, wrongly set, or buckled blades.
- Wear protective clothing and equipment suitable for the type of work.

Jigsaws

Powered jigsaws have a reciprocating blade, which moves up and down at high speed to cut timber or other material. A range of interchangeable saw blades is available.

Jigsaws are mainly used for cutting slots and curves but can also be used for straight cutting, usually with the aid of a guide attachment or fence to aid accuracy. Care must be taken to select the correct blade and speed setting, for example fast for wood, slow for metals.

Cutting can start at one edge of the material or, if cutting slots or holes, the saw blade can be inserted through a pre-drilled hole in the material to be removed.

Remember the following points.

- Always change or adjust the blades with the machine disconnected.

- Always allow the machine to run with ease, never force it around a curve.

- Select the correct blade for the task.

- Always position the machine before cutting; and avoid re-entry with an activated blade.

- Stop and allow the blade to become stationary before withdrawing it from the work piece.

- Use lubricants when cutting metals; oils for mild steel and paraffin for cutting aluminium.

- Never allow a cable to be in front of the cutter during use.

- All machines should be checked on a regular basis by a qualified electrical engineer but the operator should check the visual condition of the saw, voltage, power cable and plug.

Figure 6.16 110 V jigsaw

Sanders

There are several types of powered sander available, designed for specific tasks. The two that you are most likely to use in carpentry and joinery work are the belt and orbital sanders.

All sanders produce large quantities of dust, which can be harmful if inhaled. Always use any dust collection facility available with the machine and, if in doubt, wear a mask or respirator.

Belt sanders

Belt sanders are designed for fairly heavy-duty work and, using a coarse abrasive, can quickly remove large amounts of wood or other material like old paint. However, they can also be used to achieve a fine finish, depending on the abrasive belt used.

The fineness or coarseness of the grit used on the abrasive belt is graded according to size, as shown in Table 6.1.

Belt sanders tend to produce large amounts of dust. To compensate for this a dust bag is fitted, which collects the dust while at the same time allowing air to escape.

Figure 6.17 110 V belt sander

Grit size	Grade
20 grit and less	Very coarse
24 to 30 grit	Coarse
40 to 60 grit	Medium
80 to 150 grit	Fine
150 grit and over	Very fine

Table 6.1 Grades of abrasive belt

Figure 6.18 110 V orbital sander

Orbital sanders

Orbital sanders use abrasive sheets, held to the base of the machine by spring clips situated at the front and rear of the sander. When the machine is switched on the base moves in 3 mm-diameter orbits and, with care, a good-quality finish can be produced.

A wide range of grit sizes is available but, unlike the belt sander, they should be used for finishing only and not for removing waste material. They do not have built-in dust collection facility, so a mask or respirator should be used.

Proper tooling

Using the right cutter/bit for the right material is important, as using a wood drill bit to drill into metal or concrete will quickly blunt the drill bit's edge. However, it is equally important to know the effect of using the correct tooling.

Drill bits come in a variety of styles: masonry drill bits should be used for masonry, steel bits for metal, etc. Eventually these drill bits will become blunt and need replacing.

Sanding belts and paper used on orbital sanders will quickly become clogged with sawdust which can result in a poor finish, so these will need to be replaced.

Router cutters/planer blades and saw blades come in two different types – tungsten carbide tipped (TCT) or high speed steel (HSS). The main difference between the two is that tungsten tipped blades will keep their edges and stay sharper for longer than HSS blades. However, once a TCT blade is damaged it must be replaced, whereas HSS tooling can be sharpened.

Certain materials, for example man-made boards like plywood, will have an affect on the life of the tooling as they have been made with glue which when set becomes very hard and can dull the edge. Knots and different types of wood will also have an effect on the life of the tooling, with some woods – such as oak – being hard to work, leading to dulling of the tooling more quickly than others, say redwood.

Hazards

When using power tools there will always be dangers from saw dust or flying wood chippings. Sanding machines will create a lot of dust and saws/routers/planers will throw up wood chippings – if a loose knot is hit it can become a very dangerous projectile, so it is important that the relevant PPE is made available. This would

Safety tip

Use only belts or sheets manufactured specifically for the machine, as using makeshift belts or sheets leads to accidents

Unit 1006 Maintain and use carpentry and joinery power tools

entail goggles/dust mask /full face mask and so forth for the user and anyone in the vicinity of the machine in use.

Housekeeping is important, and it is vital that before use the area around the operation is kept clear and any debris is swept up and disposed of correctly. Sometimes, if the work creates a lot of dust/ wood chippings, the operation must be stopped *before* the debris reaches a dangerous level and the debris must be swept away before the operation can recommence.

K3. Know how to use portable power tools to drill and insert fastenings

Fixings that penetrate timber

Fixings are items of ironmongery that enable a carpenter to connect components together. When choosing a fixing we must consider certain factors, which include:

- What strength must the fixing have?
- Where will the fixing be used?
- Will the fixing need to be removed at a later date?
- Cost.

The best sources for finding out what fixings are available are trade catalogues, local builders' merchants or DIY superstores. New types of fixing are regularly added to an already extensive range.

Although there are many specialist fixings available the most common are:

- nails
- screws
- wall fixings for solid walls
- wall fixings for hollow walls
- adhesives.

Nails

Nails consist of a head and shank and are inserted by a hammer or mechanical tool. There are several types, made from either ferrous or non-ferrous metal. Ferrous metals contain iron and will therefore rust unless protected. The carpenter must decide on the most appropriate nail for the required application.

Round wire nails are available in sizes from 25 mm to 150 mm. They should not be driven below the surface of the timber and are relatively easy to remove. They are used for low-quality work where they will not be seen such as roofing and studwork.

Did you know?

Some hardwoods are acidic and when unprotected ferrous metals are inserted the process of oxidisation (rusting) is accelerated and they stain the timber

Figure 6.19 Round wire nails

Key term

Sherardising – a process of covering metal with zinc, a non-ferrous metal, to reduce rusting

Figure 6.20 Annular ring nails

Figure 6.21 Oval wire nails

Figure 6.22 Lost head nails

Figure 6.23 Panel pins

Annular ring nails are available in sizes from 20 mm to 75 mm and are also **sherardised** to prevent rusting. These nails are similar to round wire nails but feature a series of rings along the shank that makes them much harder to remove, and also provides a stronger hold.

Oval wire nails are available in sizes from 25 mm to 100 mm. They are manufactured from ferrous metal and can be punched below the surface of the timber. They are less likely to split the grain of the timber and are usually used for higher-quality work than the round wire nail.

Lost head nails are available in sizes from 40 mm to 75 mm. The head can be punched below the surface of the timber for concealment.

Panel pins are available in sizes from 20 mm to 40 mm. They are easy to punch below the surface, causing little damage to the face of the work. They are used for fine applications. Variations include sherardised and brass versions that resist rust, and veneer pins for extra fine work.

Other nails include plasterboard, felt and clout nails, also plastic-headed nails for use with uPVC systems and double-headed nails for shuttering work. All are designed for specific applications.

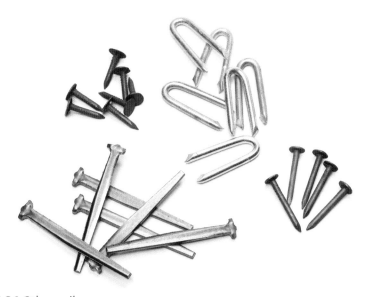

Figure 6.24 Other nails

Screws

Most modern screws are computer designed. Like a nail, screws consist of a head and a shank. However, the shank is threaded and designed to pull the fixing into the material into which it is being inserted.

Screws are manufactured from both ferrous and non-ferrous materials and are defined by:

- head type
- length, measured from the tip to the part of the head that will be flush with the work surface, ranging from 12 mm to 150 mm
- gauge (the diameter of the shank), ranging from 2 mm to 6.5 mm.

Once again, it is the carpenter's responsibility to choose the correct screw for the application in which it is being used.

Head types

Screws with countersunk heads are used when the screw has to be flush with the work or below it.

Raised head screws are usually used for attaching metal components such as door handles. Round heads are usually used for attaching sheet material to timber that is too thin to countersink.

Mirror screws have a thread within the head to which a decorative dome can be attached. As the name suggests, these are used mainly for fixing mirrors.

Pan head and flange heads are commonly found on self-tapping screws where the fixing of sheet metal is involved.

> **Remember**
>
> Screws are still sold by the old Imperial Measures gauge number (8, 10, 12, etc.), but these are being phased out

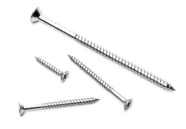

Figure 6.25 Countersunk screws

Figure 6.26 Raised head screws

Figure 6.27 Mirror screw head

Figure 6.28 Pan head

Figure 6.29 Flange head

Screwdriver head types

There are various screwdriver heads available, designed to fit each type of screw head and size of screw. A selection is shown in the following photographs:

Figure 6.30
Posidrive screwdriver head

Figure 6.31
Phillips screwdriver head

Figure 6.32
Slotted screwdriver head

Figure 6.33 Security screwdriver head

Wall fixings for solid walls

Plug fixings

Nails and screws can be used to fix components to masonry. However, the carpenter must first plug the masonry, which can either be done using a plugging chisel or an electric drill.

If a plugging chisel is used carpenters can make their own plugs, but this is time-consuming and not commonly done now. When using an electric hammer drill to plug a wall, several plug types are available, including:

- moulded plastic plugs
- hammer plugs
- frame fixings.

All of these work on the same principle. A plastic segmented sleeve fits snugly into a hole that has been pre-drilled to the plug manufacturer's stated dimensions. A screw is then inserted into the plastic sleeve that pushes the segments apart to grip the side of the hole.

Figure 6.34 Moulded plastic plug

Figure 6.35 Hammer plug

Figure 6.36 Frame fixings

Anchor bolts

These are used for giving an extra strong fixing in concrete or masonry. They consist of a segmented metal sleeve that encases a conical plug on the end of a bolt. As the bolt is turned clockwise the conical plug rises up the thread of the bolt, expanding the metal sleeve.

Figure 6.37 Anchor bolt

Wall fixings for hollow walls

Hollow wall fixings work on the principle of the fixing opening out behind the wall panel and gripping it in some way. These include:

- nylon anchors
- plastic collapsible anchors
- metal collapsible anchors
- rubber sleeve anchors
- gravity toggles
- spring toggles.

Other fixings are available called EASI drivers™, but these are generally for use only in plasterboard, although they will support heavy items such as kitchen wall units or radiators. It is worth the effort to experiment with different types of fixing before deciding which one to use.

Figure 6.38 Nylon anchors

Figure 6.39 Plastic collapsible anchors

Figure 6.40 Metal collapsible anchors

Figure 6.41 Rubber sleeve anchors

Figure 6.42 Gravity toggles

Figure 6.43 Spring toggles

Powered screwdrivers

Powered screwdrivers are a specialised simple powered drill, only available in a cordless form and specifically designed to drive or remove screws. They generally do not have removable batteries, so the whole tool is inserted into a purpose-built charger.

They are only designed to operate at slow speeds and should have variable torque control so that screws are not over-tightened. The chuck can take a full range of screwdriver bits.

Their shape enables them to be used in areas that would be difficult with a power drill.

Care needs to be taken when using powered screwdrivers, as they come with a risk of injury. One of the main injuries associated with powered screwdrivers occurs when they slip out of the screw and stab the other hand. As with using most tools, hands need to be kept out of the way and care must be taken when charging. Checks for loose or damaged wires must be carried out before the power supply is switched on.

Figure 6.44 Cordless screwdriver

Nail guns

There are three main types of nail guns in use in the construction industry today: cartridge powered nail guns, air powered nail guns and portable nail guns.

Figure 6.45 Cartridge powered nail gun

Cartridge powered nail guns are primarily used for fixing timber into concrete or steel and work by ramming a piston along the barrel of the gun at high speed. Pressure forces the nail into the materials. When the trigger is pulled, a mechanical pin fires into the cartridge, causing a reaction and forcing the piston along the barrel. Cartridges are colour-coded, allowing you to select the correct cartridge for the materials you are fixing into. The nails used are special steel nails and you must use the nails specified by the manufacturer. Manufactures may use different colour coding, so check the instructions prior to use.

Air powered nail guns work in a similar way, with a piston firing the nail into the materials, although they are generally used for timber to timber fixing. Air powered nail guns are operated by a pneumatic process and an air compressor is required to give the force needed to insert the nail. Again, special nails supplied by the manufacturer must be used and usually come in a cartridge or strip form. As these guns use a compressor for power they are mainly found in workshops and are not the preferred portable option.

Figure 6.46 Air powered nail gun

Figure 6.47 Portable nail gun

Portable nail guns also use the piston method for fixing into materials but are usually powered by gas and battery. When the gun is pushed into place the pressure allows a small amount of gas into a chamber to mix with the air. Pressing the trigger produces a spark, which causes combustion and forces the piston and the nail into the work materials. The gun comes with a charger and battery, although you are recommended to buy at least one spare battery so that you can continue work while the other charges. Nails come in strip form, which must be bought from the manufacturer. They usually come with sufficient gas cartridges to fit the amount of nails supplied. There are two main types of portable guns available: one is generally used for first fixing, and the other is a slightly smaller model used for finish work.

Nail gun safety

Nail guns are very dangerous. They cannot be fired unless they are pressed into the work. The pressure allows the trigger to be pulled. When using nail guns keep both hands away from the work area, as the nail may hit a weak spot and shoot straight through or even hit a very hard knot and shoot out at an unpredictable angle. Safety glasses – as well as the usual PPE – must be worn at all times when operating a nail gun.

Locating services

The services are specialist components within a building ranging from running water to electricity. The main services in a standard house are:

- Electrical – covers all electrical components within the building from lights to sockets. Electrical installation and maintenance work must be undertaken by a fully trained specialist as electricity can kill.
- Mechanical – covers things such as lifts. As with electrical services, work on mechanical services should only be undertaken by a specialist.
- Plumbing – can cover running water as well as gas, but only if the plumber has been recognised and qualified as a gas installation expert.

One of the main dangers associated with drilling into walls or cutting into floors is the fact that there may be a hidden service such as electric cables, water pipes or even gas pipes behind the plaster or beneath the floor boards.

There are tools available for detecting these services, and they work by checking the area for the metals found in pipes and electric wires. There are also specialist tools which can detect an electric current or the flow of water. These tools are not infallible, and if you are in any doubt that the area you are cutting/drilling into has services then it is always safest to turn off the services at the mains.

Functional skills

At the end of this unit you will have the opportunity to answer a series of questions on the material you have learnt. By answering these questions you will be practising the following functional skills:

FE 1.2.3 Read different texts and take appropriate action.

FE 1.3.1 – 1.3.5 Write clearly with a level of detail to suit the purpose.

FM 1.1.1 Identify and select mathematical procedures.

FM 1.2.1c Draw shapes.

Figure 6.48 Services locator

Working Life

Will is about to use a jigsaw. What safety checks would you advise Will to carry out before he uses the jigsaw? Can you think of any safety checks Will should carry out while he is using the jigsaw? Finally, when Will has finished with the jigsaw, is there anything you would recommend that he does?

Remember

All service work must be carried out by a fully trained and competent person

FAQ

Can I use a 110 V tool with a 230 V power supply?

Yes and no. You cannot plug a 110 V tool into a 230 V socket simply because the plug and socket are different. If you use a transformer, you can use a 110 V power tool with a 230 V power supply as the transformer will 'knock down' the supply.

Do I always have to turn off a power tool to make an adjustment?

Yes, of course. Not only must you turn the machine off, you must also make sure it is removed from the power supply. When you are making the adjustment the tool may be accidentally turned on and you could be seriously injured.

Why do I have to fully unwind a reel-up extension cable when I use one with a power tool? Surely the unwound cable could cause an accident.

An extension cable left wound around the reel can overheat and cause a fire. Unwind the cable fully but, to prevent an accident, make sure the cable is not lying in an area where people will be walking.

Check it out

1   Name three portable power tools and state a task for which each can be used.

2   What type of work is a percussion drill best suited for?

3   Explain the advantage of using a two-speed drill.

4   Why is it necessary to secure material being cut with a portable saw?

5   Explain three things that should always be checked by the operator before using a portable power tool.

6   Explain why it is important to maintain and clean power tools regularly.

7   State three points to remember when using a portable circular saw.

8   How are abrasive sanding belts graded?

9   Describe the purpose of an anchor bolt.

10   Show – using sketches if necessary – the difference between a Philips and a Posidrive screwdriver.

Getting ready for assessment

The information contained in this chapter, as well as continued practical assignments that you will carry out in your college or training centre, will help you with preparing for both your end of unit test and the diploma multiple-choice test. It will also aid you in preparing for the work that is required for the synoptic practical assignments.

The information contained within this chapter will aid you in learning how to identify various portable power tools as well as giving you information on their safe uses.

You will need to be familiar with how to:

- maintain and use portable power tools
- use portable power tools to cut, shape and finish
- use portable power tools to drill and insert fastenings.

This unit should have made you familiar with the types of power tools you will encounter in carpentry, and the purposes of each of these tools. You should be familiar now with how to use these tools safely. These skills will be vital for all carpentry work, not only in the synoptic practical assignments, but also in your professional career.

For example learning outcome 1 requires you to describe and understand the different types of power source that are used for power tools. You will also need to remember from Unit 1001 how to work safely with electricity. You will need to apply this knowledge of power sources to working with the different power tools and to charge batteries correctly. This unit has also explained the importance of having valid PAT date certificates. When working with power tools you will need to check this as well as the tools themselves, and their cables, for damage.

This unit has also explained the different types of tooling and their uses. You will need to be familiar with these, and understand the difference between them, as you will need to be able to change these tools depending on the particular task you need to work on.

Perhaps the most important fact to remember from this unit, is the importance of storing and maintaining power tools correctly and safely, and following current legislation on their use. This knowledge will be vital at any time when you are using power tools, whether this is during the practical assignment or in your professional career.

The speed and accuracy with which you use these tools will also help you to prepare for the time limit and the quality of work that will be required to pass the synoptic practical task. But remember, do not try to rush the job as speed will come with practice and it is important that you get the quality of workmanship right. You will also need to be sure that you are using the tools safely.

Before you start work on the synoptic practical test it is important that you have had sufficient practice and that you feel that you are capable of passing. On approaching the test it is best to have a plan of action and a work method that will help you as well as a copy of the required standards, associated drawings and sufficient tools and materials. It is also wise to check your work at regular intervals. This will help you to be sure that you are working correctly, and help you to avoid problems developing as you work.

Always make sure that you are working safely throughout the test. Make sure you are working to all the safety requirements given throughout the test and wear all appropriate personal protective equipment. When using tools, make sure you are using them correctly and safely.

Good luck!

CHECK YOUR KNOWLEDGE

1 What is the minimum age to use a power tool?
- **a** 16
- **b** 17
- **c** 18
- **d** any age, as long as you have been trained.

2 What is the correct power tool voltage on site?
- **a** 55 V.
- **b** 110 V.
- **c** 220 V.
- **d** 240 V.

3 What is the maximum amount you can remove with one pass of a portable electric planer?
- **a** 2 mm.
- **b** 3 mm.
- **c** 4 mm.
- **d** 5 mm.

4 Which PPE must be worn when using a portable electric planer?
- **a** goggles and boots.
- **b** ear defenders and gloves.
- **c** dust mask.
- **d** all of the above.

5 What operations can a circular saw perform?
- **a** ripping.
- **b** cross-cutting.
- **c** bevel cutting.
- **d** all of the above.

6 A chop saw can cut:
- **a** mitres.
- **b** square.
- **c** compound.
- **d** all of the above.

7 What is the safest voltage to use on site?
- **a** 55 V.
- **b** 12 V.
- **c** 110 V.
- **d** 240 V.

8 Cordless tools are better because:
- **a** there is no cable, so they are more portable.
- **b** the battery runs flat quickly.
- **c** they are cheaper.
- **d** they can't go in reverse.

9 Which sander should be used for fine finish work?
- **a** belt sander with fine sandpaper.
- **b** orbital sander with fine sandpaper.
- **c** belt sander with coarse sandpaper.
- **d** orbital sander with coarse sandpaper.

10 What is one of the main dangers of drilling into walls?
- **a** hitting hidden services.
- **b** damaging the plaster.
- **c** flying debris.
- **d** drilling your hand.

Index

A-frames 56
accident book 15, 17
accidents 15-20, 54
    cost of 18-19
    fatal 1, 6, 16, 17, 18
    making area safe 20
    near misses 17-18, 205
    reporting 4, 6, 15-16, 205
addition calculations 89-90
adhesives 47, 156
aggregates 41-3
airbag safety system 51-2
alcohol, drinking 27
anchor bolts 220, 221
angled joints 155
axes 198-9

bar charts 85-7
battens 130
belt sander 215
bits
    drill 192-4, 210, 216
    router 211
block plans 82
blocks: storage and use 39
blockwork walls 125
body language 111, 112-13
bonding 124
bracing 131
bradawl 191, 192
bricks: storage and use 38-9
brickwork 127
    bonding 124
bridle joints 155
British Standards Institute 52, 80
burns 24
butt and edge joints 155

calculations 87-92
callipers 142

carcassing timber 44-5
carpenter's brace 191, 192
cavity walls 125
CDM Regulations (2007) 8
cement 43-4
cement-sand plaster 43
chemical spills 23
chemicals, storage of 21
chipboard 170
chisels 196-7
    sharpening 179-80
clamps 195-6
combination square 140
communication 111-13
    positive and negative 110
    see also information
compressed air powered tools
    36, 38, 222
concrete 42
concrete floors 123
confidentiality clauses 76
confirmation notice 106
Construction Skills
    Certification
    Scheme (CSCS) 11, 14
contingency plans 87
cordless tools 206, 221
COSHH Regulations 7, 23,
    29-30
crawling board 55
cutting gauge 142
cutting lists 138

daily report/site diary 109
damp proof course 122
Data Protection Act (1998) 76
datum points 118-20
day worksheet 105
decimal calculations 90
decorating materials storage
    47-8, 49

delivery notes 108
delivery records 109
diseases 1, 6, 16-17, 31-2
division of numbers 91-2
documents
    security 76-7
    site paperwork 104-9
double insulation 206
dovetail joints 152-3
drawings 77, 137
    abbreviations 77, 81
    scales 77, 78-9
    symbols 79-81
drills
    bits 192-4, 210, 216
    hand-held 191-4
    power 208-10
drugs 27-8

ear damage and noise 28-9
ear-plugs/ear defenders 29, 66
electric shock
    first aid 63
    precautions 62-3, 205
electrical tools 36-7
Electricity at Work
    Regulations (1989) 8
electricity, working with 61-4
    hazards 24
emergencies 14-15, 16
    fire 68-70
emergency services 16
employees' duties 2, 4, 8, 10, 50
employers' duties 2, 4, 8, 10, 50
English bond wall 124
equipment 9, 36-61
    see also tools
eye protection 65-6, 204

fall-arrest systems 50-2
falling from height 23

Index

fibreboard 170-1
files 190, 191
fire 15, 24, 68
    evacuation procedures 70
    legislation 11
    muster points 70
    triangle of 68
fire blanket 70
fire extinguishers 68-70
fire-fighting equipment 68-70
firring 127, 129
first aid 19
    electric shock 63
    Regulations 11
first aid box 19
first aiders 16, 19, 20
fixings 217-21
    for hollow walls 221
    for solid walls 220
flammable substances 3, 21-2
    Regulations 11
flashings 131
flat roofs 127-8
Flemish bond wall 124
floating floor 123
floor plan 84, 85
floors 122-3
foot protection 66
foundations 121-2
frames 56-7

Gantt chart 86-7
gas powered tools 36, 38
gauge box 126
gauges 140
glass storage 22
gouges 197-8
ground floors 122-3
guidance notes, health and
    safety 2
gypsum plaster 43

halving joints 155
hammers 143-4

hand gestures 112
hand protection 67
hand tools 37
    maintenance 179-86
    safety rules 178
hand washing 26
handsaws see saws
hardboard 171
hardwoods 45, 152, 159, 163-4
    identification and uses 166-8
hazard books 26
hazardous substances 7, 29-30
    decorating materials 49
    risk assessment 30
    storage 21-2
    waste 31
hazards on construction sites
    20-4
head protection 65
health and hygiene 26-32
health risks in workplace 31-2
health and safety
    policy 110
    proactive approach 2, 18
    'reasonably practicable'
    phrase 2, 3
    regulations 2-11
    sources of information 11-12
    tools and equipment 36-61
Health and Safety at Work Act
    (1974) 3-6, 204
Health and Safety Executive
    (HSE) 2, 5-8, 11
    on PPE 65
    reporting accidents 6, 16-17
health surveillance 7, 30
health and welfare 15-20
hearing protection 7, 29, 66
height see work at height
high-visibility jackets 67
hop-ups 57
housekeeping 20-2, 23
    and accidents 23
    hazardous substances 21-2

imperial measurements 94
induction, site 12-13
infections 17, 31
information 103
    communicating 75-116
    document security 76-7
    site paperwork 104-9
    types of document 76
injuries 1, 6, 16-17
inspectors, HSE 5-6
insulation board (softboard)
    171
internal walls 125
invoices 108

jig saws 214-15
job sheet 105
jointer plane 187
joints 135-75
    dovetail 152-3
    lengthening 154
    mortise and tenon 148-52,
    154
joist, upper floors 123

kerbs 40
kiln seasoning 161, 162-3
kinetic lifting 32-4
kitemark, BSI 52, 206
knots in timber 158-9

ladders 53-5
    erecting 54-5
    roof 55
laminated boards 169-70
laser levels 143
legislation
    fire 11
    health and safety 2-11
    list of Acts 11
    power tools 204
    waste materials 30, 31
lengthening joints 154
leptospirosis 31

levelling tools 142-3
lever tools 147
lime-sand plaster 43
lintels 126
locating services 223
location drawings 82-3
LPG (liquefied petroleum gas) 21-2
Regulations 11, 22
lung diseases 16, 31

mallets 144
Management of Health and Safety at
Work Regulations (1999) 24
manual handling 9-10, 32-5
kinetic lifting 32-4
Regulations (1992) 9-10, 32
marking knife 140
marking out 136-43
cutting list 138
setting out rod 136-7
tools 139-47
masonry, internal and external 124-7
materials 47-8, 49
estimating quantities 87-92
weather protection 38, 39, 40, 48
measurements 92-102
area of circle 101-2
circumference of circle 100-1
composite shapes 99-100, 102
perimeters 94-5
Pythagoras' theorem 97-9
units of 92-3
medium density fibreboard (MDF) 171
message taking 103
method statements 25-6
metric measurements 93
mobile tower scaffolds 59-60
mortar, mixing 126-7

by hand 126
by machine 127
mortise gauge 140
mortise and tenon joints 148-52, 154
multiplication 90-1

nail guns 222-1
nails 217-18
narrow strip foundation 121
near miss reporting 17-18, 205
noise 28-9
Regulations (2005) 7
noise induced hearing loss 28-9
numbers, positive and negative 88-92

oilstone 179
orbital sanders 216
orders/requisitions 107
Ordnance bench mark (OBM) 118-19, 120

painting materials storage 47-8, 49
PAT (portable appliance testing) 36, 207
paving slabs 39-40
penalty clause 87
pencils 140
percussion tools 143-5
pitched roofs 128-9
planes 187-90
powered 210
sharpening 179-80
using hand-held 187-90
plaster 43-4
plasterboard 44
platforms 57
pliers 194
plug fixings 220
plywood 45, 169
policies and procedures 110

portable circular saws 212-13
powdered materials 48, 49
power drills 208-10
power supplies 206
power tools 36-7, 203-26
hazards 216-17
maintenance 207-8
proper tooling 216
safety issues 204-5
storage 37-8
powered planers 210
PPE (personal protective equipment) 4, 64-7
for handling tools 36
power tools 205, 210, 212, 213, 216–17
Regulations 10, 65
storage and maintenance 64-5
types of PPE 65-7
progressive kiln 153
proprietary tower 59
punches 144-5
purlin 129
PUWER (Provision and Use of Work Equipment Regulations) (1998) 9, 36, 204, 207
Pythagoras' theorem 97-9

raft foundation 122
range drawing 84
rasps 190, 191
ratios: scale drawings 78
RCD (residual current device) 25, 62
record keeping 76-7
respiratory protection 66, 170, 215, 217
RIDDOR (Reporting of Injuries, Diseases and Dangerous Occurrences Regulations) (1995) 7, 15, 16

rip saw 182, 183
risk assessment 4, 26
 five steps 24-5
 hazardous substances 30
 manual handling 9-10
 working at height 49-50
rolled materials, storage of 41
roof construction 55, 127-31
 felt, slate and tiles 131
 finishing roof 129
 prefabricated truss roof 128
 roof components 129-31
roofing tiles 40-1, 131
routers 211-12, 216
Royal Society for the
 Prevention of
 Accidents (RoSPA) 12
Royal Society for the
 Promotion of Health
 (RSPH) 12
rules, folding 139

safe working practices 1-74
 see also health and safety
safety netting 51
safety officers 16
safety policy 2, 4
safety signs 30, 71
sand, types of 42
sanders 215-16
saws 186
 chop saws 213-14
 jig saw 214-15
 portable circular 212-13
 setting and sharpening
  180-2
 types of 183-6
scaffold boards 57
scaffolding 58-61
 types of fittings 59
scale drawings 77, 78-9
schedules 84-5
screwdriver bits 193
screwdriver heads 219

screwdrivers 145-7
 powered 221
screws 218-19
seasoning of timber 161-3
second seasoning 47
security alerts 15
services, locating 223
setting out rod 136-7
setting out tools 136
sheet materials storage 44-5
signs, safety 71
site 4, 110
 hazards on 20-4
 induction 12-13
 paperwork 104-9
site datum 118-20
site plans 82-3
skin diseases 16, 31
skin and sun protection 67
sliding bevel 140
smoke inhalation 24
softwoods 152, 159, 163-4
 identification and uses 164-5
soil, surveying 121
specifications 83-4
spillages 23
spokeshaves 190
staging boards 57
staircase joints 153-4
steel trestles 56-7
stepladders 52-3
stepped shoulder joint 151
stopped housing joint 153
storage of materials 46-8, 49
stretcher bond wall 124
stud partition 125
subcontractors 3, 4
substance abuse 27-8
substructure 120-2
subtracting numbers 90
suppliers: health and safety 3,
 4-5
surforms® 190, 191
suspended timber floor 122-3

temporary bench mark
 (TBM) 120
tensile forces 153
timber 163-4
 carcassing 44-5
 conversion 159-61
 cutting list 138
 defects 156-7, 158-9
 floors 122-3
 identification of 164-8
 seasoning 161-3
 storage 45-7
timber manufactured boards
 163, 169-71
timber stud partition 125
timesheet 104
toolbox talks 13-14
tools 36-61, 139-47
 chisels 196-9
 drills 191-4
 hand tools 179-86
 holding and clamping 194-6
 maintenance and use 177-
  202
 marking out 139-43
 planes 187-90
 power tools 36-7, 203-26
 protection 182-3
 safe storage 37-8
 saws 183-6
 shaping 190-1
tower scaffolds 60-1
 low towers 60
trestle platforms 56-7
tri-square 140
tripping accidents 23
twin tenon with twin haunch
 joint 151
twist bits 182, 183

upper floors 123

variation order 106
verbal communication 111

vibration white finger 32
volatile substances 49
voltages 62
    power tools 205, 206

wall plates 130
wall ties 125
walls 125-6

waste, hazardous 31
    legislation 30, 31
welfare facilities 27
wheelbarrows 38
wide strip foundation 121
wood *see* timber
Work at Height Regulations
    (2005) 10

work programme 85-7
working at height 23, 49
    dangers of 49-50, 61
    fall protection 50-2
    risk assessment 49-50
working platforms 49-61
written communication 111